FRENCH MILITARY AVIATION

MILITARY HISTORY BIBLIOGRAPHIES
(General Editors: Robin Higham, Jacob Kipp)
Vol. 11

GARLAND REFERENCE LIBRARY
OF THE HUMANITIES
Vol. 381

MILITARY HISTORY BIBLIOGRAPHIES

Advisory Editors:
Robin Higham
Jacob W. Kipp

1. *Israeli Military History: A Guide to the Sources*
   by Jehuda L. Wallach

2. *German Military Aviation: A Guide to the Literature*
   by Edward H. Homze

3. *Germany Military History: 1648–1982: A Critical Bibliography*
   by Dennis E. Showalter

4. *The Military in Imperial History: The French Connection*
   by Alf Andrew Heggoy and John M. Haar

5. *Japanese Military History: A Guide to the Literature*
   by Shinji Kondo

6. *French Military History, 1661–1799: A Guide to the Literature*
   by Steven T. Ross

7. *German Naval History: A Guide to the Literature*
   by Keith W. Bird

8. *Balkan Military History: A Bibliography*
   by John E. Jessup

9. *Napoleonic Military History: A Bibliography*
   edited by Donald D. Howard

10. *British Military History: A Supplement to Robin Higham's* Guide to the Sources
    by Gerald Jordan

11. *French Military Aviation: A Bibliographical Guide*
    by Charles Christienne, Patrick Facon, Patrice Buffotot, and Lee Kennett

FRENCH MILITARY AVIATION
A BIBLIOGRAPHICAL GUIDE

Charles Christienne
Patrick Facon
Patrice Buffotot
Lee Kennett

Garland Publishing, Inc. • New York & London
1989

Library of Congress Cataloging-in-Publication Data

French military aviation.

(Military history bibliographies; vol. 11) (Garland reference library of the humanities; vol. 381)
Includes index.
ISBN 0-8240-8518-3 (alk. paper)
1. Aeronautics, Military—France—History Bibliography. I. Christienne, Charles. II. Series. III. Series: Garland reference library of the humanities; vol. 381.
Z2181.M5F73 1989 016.3584'00944 88-33470
[UG635.F8]

Printed on acid-free, 250-year-life paper
Manufactured in the United States of America

TABLE OF CONTENTS

General Editor's Introduction.....vii
Preface.....xi

PART I—HISTORICAL PROBLEMS AND PERSPECTIVES

The Beginnings, 1783–1870.....1
Developments from 1871 to 1914.....6
The Great War.....21
Military Aviation from 1919 to 1939.....38
The Second World War.....60
The Armée de l'Air Since 1945.....79

PART II—BIBLIOGRAPHY

Guide to Abbreviations of Periodicals.....85
Guide to Research Materials.....86
The Beginnings, 1783–1914.....103
The First World War.....112
The Interwar Period.....131
The Second World War.....195
The Period Since 1945.....217

THE GENERAL EDITORS' INTRODUCTION

Military history is a vast international field touching upon all aspects of society, yet guides to the immense literature it contains, essential for anyone writing in the field, have largely been conspicuous by their absence. Over fifteen years ago an international start was made by Robin Higham with *Official Histories* (1970), which for the first time enabled even official historians to check what other offices had produced. That work established a pattern of historical essays, some by the original authors, followed by a checklist of the works published. While this was in process, *A Guide to the Sources of British Military History* (1971) was commenced along slightly different lines. In this each author was asked to provide a bibliographic essay over a segment of the military, naval or air history of one country, with a numbered alphabetical list by authors of those cited following. A further feature of each chapter was a short concluding section suggesting what research remained to be undertaken as well as what could profitably be reviewed. So successful was this volume, published simultaneously in the United States and in Britain, that not only did it become a standard library reference work, but also the editor was asked to undertake a similar compilation for U.S. military history. This latter guide was completed in 1975 on a more far-seeing scheme in which quinquennial supplements were planned to keep the work up to date (the first of these was issued in 1981). It was on the basis of these three successful volumes that the editors were approached by Garland in May 1978 to undertake the present series.

By that time both editors were involved in the United States and in the International Commissions for Military History, Dr. Higham being on the Editorial Advisory Board and Dr. Kipp on the Bibliographical Committee. They agreed with Garland that they would undertake to produce an international series of some thirty-odd volumes to provide scholars access to the vast collections in all the countries in the world other than the United States and the United Kingdom. Those authors whose native language was not English should also produce a volume in their own tongue so that scholars of at least two major languages concerned with the military history of that country should benefit. No limits were set on length, though as with the two previous *Guides* it was not expected that these works would be either totally comprehensive or exhaustive, if for no other than the two very good reasons that in most cases there is a lot of repetitive material and that inevitably bibliographies are dated from the moment their sections are completed.

The very existence of the International Commission on Military History and the vigorous programs of the various national commissions that compose it confirm the vitality of the field of military history. Both editors believe that this bibliographic series can render a valuable service by facilitating the study of military history across national lines and in a comparative context. It is hoped that each volume will make explicit the general and particular historiographic approaches associated with an individual nation's military institutions. In such a manner researchers will be able to consult a single volume that will outline the historigraphic developments for a particular nation's military experience or, in the case of major powers, of that of one of its principal armed services. Such a guide, it is hoped, will serve as a compass for further research, both enumerating what has been done and suggesting what still needs to be done.

Essentially the authors have been asked to provide readable essays that will guide the readers through the labyrinth of the most important sources as though walking through a library and archives, mentioning authors and titles, but leaving the details to be acquired by taking the works from the shelves, in this case the numbered lists at the back of each chapter. Both authors and editors would be happy to have comments and suggestions from readers and users.

Jacob W. Kipp
Soviet Army Studies Office
U.S. Army Combined
Arms Center
Fort Leavenworth, KS 66027

Robin Higham
Department of History
Kansas State University
Manhattan, KS 66506,USA

PREFACE

This book is intended as a guide and an aid to historical researchers in the field of French military aviation. It is divided into two basic parts, the bibliography proper being preceded by an extended treatise on the development of military aviation in France and the challenges and opportunities that subject offers to historians. In this essay General Christienne has sought to indicate the major gaps in historical knowledge, the problems of filling them, and the lines of future inquiry which seem most promising.

The bibliographical section has a prefatory essay which surveys briefly the wide range of source materials the historian has at his disposal, the repositories that would be most useful, and the general and background works on the subject. Subsequent sections are arranged chronologically, with a brief bibliographical essay preceding each bibliographical listing. In all some two thousand books, articles, dissertations, and theses are listed, the great bulk of them being understandably in French. While this compilation makes no claim to completeness, every effort has been made to include significant and representative publications through 1985.

Charles Christienne
Patrick Facon
Patrice Buffotot
Lee Kennett

PART I

HISTORICAL PROBLEMS AND PERSPECTIVES

THE BEGINNINGS, 1783-1870

LES AEROSTIERS DE LA REPUBLIQUE (THE BALLOONISTS OF THE REPUBLIC)

It was in France, during the Revolution of 1789, that the air weapon was first used in war. A student of the history of aviation can easily understand how this first employment should have occurred in that era of intellectual, ideological and scientific ferment which marked the end of the Enlightenment. And it was in France that the birth and development of lighter-than-air craft began when towards the end of 1783 two famous firsts occurred there: on November 21 the ascension of Pilâtre de Rosier and the Marquis d'Arlandes in a hot air "montgolfière," and on December 1 the flight of Jacques Charles and Nicolas-Louis Robert in a hydrogen balloon. One year later, January 7, 1785, Jean-Pierre Blanchard left Dover in a hydrogen balloon in company with an American named Benjamin Jeffries; they crossed the Channel, demonstrating a fact not fully understood at the time: England was no longer an island. The enthusiasm which greeted these early flights was considerable, evidenced by the appearance of the balloon as a popular symbol in decoration.

From the very first experiments the military possibilities of the aerostats were apparent, and the Marquis d'Arlandes discussed them in the report of his first ascension. But it was the celebrated Committee of Public Safety, created on April 18, 1793, that examined all scientific developments with a view to enhancing the nation's war effort, that decided on the use of military balloons. A team made up of the savants Louis-Bernard Guyton de Morveau, Antoine-Laurent Lavoisier, and Jean-Marie-Joseph Coutelle developed a new process for making hydrogen by passing water over iron filings heated red hot in a brick furnace. The construction of these furnaces was the first task of the company of AEROSTIERS or balloonists created on the 13th of Germinal, in the Year II (April 2, 1794). This also was one of the reasons why the company was attached to the engineers, and explains why that branch of the service long kept military aviation under its purview.

On the 14th of Floréal, Year II (May 3, 1794), the company at full strength joined the Army of the Sambre et Meuse, commanded by General Jean-Baptiste Jourdan. It had as its purpose to place at the general's disposal all the means which the aeronautical art could supply. Its mission as outlined in the somewhat revolutionary language of the official instruction, was threefold:

1. to reveal the marches, movements and plans of the enemy,
2. to transmit speedily signals agreed upon in advance with generals of division and commanders of fortified places,
3. to reply, as circumstances might permit, to messages coming from regions occupied by satellites of the despots.

From the time they entered the campaign, the balloons rendered the services which had been expected of them, first at the siege of Maubeuge, then at the battle of Fleurus. Encouraged by these successes, the Committee of Public Safety decided to create a second company, and on the 10th of Brumaire, Year III (October 31, 1794), the Committee created at Meudon a National Balloon School, under the direction of Nicolas Conté.

We will not trace in detail the further activities of the two companies, but only note that eventually their usefulness came to be challenged. This challenge is interesting because it reappears in similar form in the Union Army during the American Civil War. When positional warfare was replaced by a war of movement the use of balloons became more difficult; from that point it was easy to say that they were of no utility, and on the 28th of Pluviôse, Year VII (February 18, 1799), the companies were dissolved and the school at Meudon abolished. Were the arguments put forth to justify this abolition valid, or were there hidden resentments and less noble motives behind the decision? General Pierre Lissarague, former Director of the Musée de l'Air et de l'Espace, believes that there were such motivations and that already the renown of the aeronauts was beginning to overshadow that of the generals. After all, what would become of the general's prestige as he stood on his hilltop with his spyglass, trying to incarnate military genius, if a young lieutenant, carried aloft by an envelope of hydrogen, could see the enemy's movements and perhaps inspire the countermovements of the army in which he served? This sort of reaction was normal enough, and in later times it would inspire in many countries decisions catastrophic for military aviation and for the military generally.

In the final analysis this episode of the Balloon Corps of the Republic appears to be of little practical consequence, but we might note here certain thematic elements which will reappear in the subsequent evolution of military aviation. The first element is the close relationship between aeronautics and science. Here more than in any other arm of the service military innovation followed technological innovation. The first military use of the balloon follows within thirty years the discovery of the technique for making hydrogen. But military innovation can also yield technical innovation: the desire to employ balloons in the simplest manner led the scientist to devise a new technique for the production of hydrogen.

A second element is constituted by the hostility which military conservatism showed towards the air weapon virtually from the beginning. We have tried to explain this conservatism, which consigned the balloons of Fleurus to oblivion. There are other examples which would furnish a rich field of research in France and in other countries as well. At the same time it is interesting to note that this military conservatism must give way to the will manifested by a strong and revolutionary political power. It was the Committee of Public Safety, a truly revolutionary government, which decided to undertake the research leading to establishment of military aerostat units, and it was the Committee which decided the constitution of those units. Likewise it was the Committee which defined the mission of the Balloon Corps, and it is not surprising that those missions were framed in terms which inveighed against "despots."

There are very few serious studies on this subject, nor is it clear that they could be carried to fruition. Yet it would be a fascinating undertaking for a researcher to focus on the perceptions, the viewpoints of the first Balloonists, to determine what place they occupied, what considerations sustained them and their work, and what other considerations opposed them. It would be equally fascinating to examine the arguments which led to the dissolution of the Balloon Corps and to speculate on the consequences of that dissolution. We might ask ourselves the question which Liddell Hart raised: What might have happened at Waterloo if Napoleon had been able to profit from the services of the Balloon Corps?

FROM 1800 TO 1870

The memory of the balloons of Fleurus and Mayence lingered in popular memory thanks to paintings and engravings which portrayed those feats of arms -- but they inspired few reflections among French military leaders. While it is true that in 1826 the Minister of War named a commission to study the possibilities of aerostation, its report was never implemented. Until 1870 the few attempts to use balloons for military purposes were carried out on the initiative of civilians, as was the case of the civilian balloonist Margat in 1830 in the course of the expedition to Algeria, and with Eugène Godard during Napoleon III's Italian campaign of 1859.

During this time civilian aeronauts continued their ascensions, amassing a considerable fund of experience. Among these many balloonists, whom we cannot treat in detail, one in particular merits study: Nadar was an artist and photographer who was also a passionate devotee of flight, convinced that the future lay in heavier-than-air craft. Along with friends he created the Society for the Encouragement of Aerial Navigation and launched in 1863 the Journal L'AERONAUTE, which proved an excellent vehicle for popularizing aeronautics as a sport.

Nadar was also among the first to experiment with taking photographs from balloons, but his first satisfactory photographs date only from 1868. By that time the American Thaddeus Lowe was already employing aerial photography as a means of observation in war. Indeed it was in Civil War America that the use of the air arm once again made its appearance. Two companies of specialists created by Lowe at the request of President Lincoln put into operation some ten balloons. Lowe did much to perfect the techniques of military ballooning; one of his most important contributions was the development of cylinders to store the hydrogen under pressure, these containers being transported by horse and wagon. This innovation freed the service from direct dependence on fixed generating ovens.

The balloons were used from 1861 to 1863, when the return to warfare with greater movement made them less easy to use. The balloon units were abolished at the end of the war. It would be interesting once again to follow the arguments put forth here. At the beginning of 1914 we find in the French high command this same persistent notion that balloons could not follow the rapid movements of armies in the field, a notion which led to aerostats being assigned only to fortified places. Thereafter French leaders were surprised to find that the Drachen balloons of the German Army were able to follow troop advances without great difficulty and to reconnoiter effectively their intended march routes. Here again we may raise the subject of military conservatism and pose the question of motive in the neglect of a technique at the very time it had demonstrated its usefulness.

The use of ballons during the Civil War raises other questions. It would be interesting to know how familiar Lowe was with the experience of the Balloonists of the Republic, and to what degree he might have drawn inspiration from them. It would also be worthwhile to try to gauge the influence which this American experience had on French military thought, and particularly on the work of Captain (later Colonel) Charles Renard, undertaken at Chalais-Meudon from 1874 on.

Certain it is that at the outbreak of the Franco-Prussian War in July, 1870 the French Army had not only made no provision for the use of balloons, but even declined the offer of two civilian aeronauts, Gaston Tissandier and Wilfred De Fonvielle, who volunteered to follow the Army of the Rhine with a captive balloon. In the course of the conflict the usefulness of the balloon was once again discovered, though at first not so much as a means of observation, but as a means of communication between besieged places and territory held by friendly forces. The first experiments were made at Metz with small balloons which carried messages, but it was above all during the siege of Paris that the French made what was for the period an extensive use of large free balloons for the transportation of passengers and mail, at the rate of three or four flights per week. One of the most celebrated passengers was Léon Gambetta, Minister of War and Interior. Designated to head the governmental delegation at Tours, he left Paris on October 7, 1870, in order to direct the war effort in the rest of the country. The unusual method of his departure, added to his revolutionary ideas and his ardent patriotism, drove him to exploit every possible means of gaining an advantage over the enemy and led him quite naturally to create once again a company of balloonists to support to operations of the Army of the Loire. Unfortunately the armistice concluded on January 28, 1871 gave the Balloon Corps no chance to prove itself. Even so, the potential which balloons had demonstrated, both in the siege of Paris and in their brief use by the Army of the Loire, would not be forgotten; aerostation would henceforth find its place in the structure of national defense, which was drastically reorganized in 1872.

DEVELOPMENTS FROM 1871 to 1914

MILITARY PROBLEMS AFTER 1871

New solutions were needed to resolve postwar military problems, which were various, but were all born from defeat. They had to be addressed without delay and in a non-partisan spirit, even if on some issues there were sharp differences among various political currents. These differences were first of all over the nature of military service. The Radicals, noting the defeats suffered by the Imperial armies and the relative successes of the Army of the Loire, and influenced by the legend of the LEVEE EN MASSE of 1793, sought the suppression of standing armies. This was manifestly not possible, so the radical program substituted the concept of the nation in arms.

The Law of August 17, 1872 was voted by a chamber with a monarchist majority and strong military representation. Having laid down the principle that every Frenchman owed military service, the law then granted numerous exemptions: elder brothers of orphans, only sons, seminary students, and teachers. The system of drawing lots was maintained and those with "bad" numbers served five years or only one year in the case of educated persons with diplomas. This law, incorporating obvious inequalities, did not provide the Army with the effectives it needed. The Law of July 17, 1889 introduced three years of service and was regarded as the first regulation introducing equality of obligation. If in fact all Frenchmen of service age were now called, those categories previously exempt served only one year. The Law of 1905 was the first to introduce absolute equality of obligation to military service. The period of active service was reduced to two years and all exemptions abolished.

In 1911 the Socialists presented through Jean Jaurès a bill which would replace the army with a militia similar to that found in Switzerland. This proposal was an interesting one from several points of view, but the General Staff opposed it. War seemed not far off and in 1913 the Parliament, accepting the views of the military, extended the period of active service to three years. This was the only means of ensuring that there would be enough troops under arms to face the German Army in the opening days of a conflict.

If the laws governing the nature of military service were hotly debated, those on Army organization and administration were voted without difficulty. The law on

Army organization adopted in 1873 divided that body into Army Corps, each to be stationed in peacetime within an Army Corps Region, this being the first application of the later Military Regions. The General commanding the Region controlled both the troops and services in his area and prepared the mobilization of units which would complete his Corps in wartime. In practice this organization proved to be excellent and it served as the basis for all military organization up to recent times. We shall see that it was chosen as organizational model for the independent air force, a choice which proved totally unsatisfactory.

In the same fashion the law of 1882 defining the rules of military administration laid down certain general principles of military organization: organizational permanence, unity of command. These principles, eminently valid for land forces, penetrated the military mentality and imposed themselves at the birth of the Armée de l'Air in 1933. It took more than twenty years for it to become apparent that such principles were not applicable to the air arm and that the distinctive nature of that arm called for different solutions. Still in all, these laws were excellent for the era and the place where they were applied. They provided a framework highly suitable for the renovation of the military. This framework was completed by the creation of the Ecole de Guerre in 1878, the establishment of the General Staff in 1890, and the reorganization of the Counseil Supérieur de la Guerre, which took on its definitive form in 1888.

This military activity was not limited to the defining of structures; it corresponded to a stirring period of activity in all units and all services, each striving to improve its contribution to the nation's military might. De Gaulle would later write: "There was not a tool, a caisson, a boat, for which some new model was not adopted between 1875 and 1890." This was notably the case with armament. In 1871 the Chassepot rifle was replaced by the Gras, which in turn gave way in 1886 to the Lebel, modified in 1893. The artillery developed new field pieces, the Bange 80 mm and 90 mm models. Siege artillery was perfected as were the 270 mm mortar and the 120 and 155 mm pieces; and lastly the 75 mm, capable of firing twenty rounds a minute, represented a genuine technical revolution. At the same time the engineer Eugène Turpin invented a very powerful explosive derived from nitric acid -- Turpinite, later renamed Melinite in honor of the Premier.

CLEMENT ADER

It is not surprising that in this atmosphere Louis-Charles Freycinet, Premier and Minister of War, should be taken with the works of Clément Ader, or that there should be intense activity at Chalais Meudon, supervised by the imposing figure of Captain Charles Renard. Ader is interesting for several reasons. Born in Muret, in the south of France, in 1841, he was the archetype of the self-made man. Of modest artisanal background, he began his career in the Ponts et Chaussées (Highway Department). On his first job he demonstrated an inventive mind and an exceptional gift for mechanics. He invented new forms for bridges and in 1866 he patented a rail-handling machine. During the remainder of his life he took out countless other patents on the most diverse matters, ranging from the rubber-tired bicycle to the hovercraft, passing by way of the "theatrophone" and the "endless rail," ancestor of the caterpillar track.

From his youth Ader was fascinated by the idea of flight. At Castelnaudary he had conducted experiments with a tethered glider and had long observed the flight of birds and insects. All of his life he believed that if man were to fly he would have to imitate as closely as possible the flight of birds. But to develop such heavier-than-air craft Ader would need a considerable sum of money. To earn this money he left government service to exploit patents which he had taken out on the telephone. He was one of the creators of the first telephone system installed in Paris, an enterprise which made him a fortune, further augmented by the sixty-odd patents registered in his name in the period from 1879 to 1904, not only in France but also abroad.

Having become a rich man, Ader could at last realize his dream. In 1882 he built the EOLE, a machine equipped with a 20 h.p. steam engine. On October 9, 1890, in the park of the Chateau of Armainvilliers, placed at his disposal by Madame Pereire, he succeeded in leaving the ground for a distance of fifty meters. This was the first time anywhere in the world that such a feat was accomplished.

From the outset -- and this is perhaps the most remarkable aspect of his work -- Ader had understood the great military advantages offered by airborne vehicles. Convinced of the military value of his invention, he succeeded in convincing Freycinet, then Premier and Minister of War to finance, at least in part, the research and development for a new prototype. To defray the expenses the

Minister had available part of the "Giffard Bequest." The legacy had been left to the government to support research in aeronautics. At that time most of the research was directed toward lighter-than-air craft. Some partisans of aerostats were angry at this "malversation" of which they felt themselves the victims. There is some reason to wonder whether the resentment which defenders of the lighter-than-air so long harbored against Ader did not have its origin in his use of the Giffard Bequest.

Ader went back to work and by the summer of 1897 he was ready to try his "Avion no. 3." Before a commission presided over by General Mensier, Ader attempted a test flight, using a circular runway, and despite a strong wind, the machine's wheels left the ground -- otherwise its brusque swerve off course cannot be explained -- then it struck the ground violently and was shattered. Despite this accident the General's report on the test was very favorable to continuing the experiments; but the Minister of War, General Jean-Baptiste Billot, decided to cut off the funds. Ader, who had sunk much of his personal fortune into the work, was profoundly discouraged; he gave up the construction of airplanes.

Here we should mention the quarrel which developed much later and which cast doubt on the authenticity of Ader's flights. In 1906 Ernest Archdeacon, President of the Aeroclub of France, was anxious to prove that Santos Dumont, that most "Parisian" of Brazilians, was the first to have taken off in an aircraft relying only on the strength of its motor. He affirmed that the wheels of Ader's craft had never left the ground. Despite the stir it made, this controversy is of secondary importance. Also secondary in interest is the rather chauvinistic argument which accords a "first" in this domain to the Wright brothers. In truth comparisons are not easy to make.

Ader is indisputably the first to have left the ground for a distance of fifty meters. No one challenges the flight of the EOLE in 1890; but it is also true that Ader's invention brought no proof that the use of heavier-than-air craft was possible in war. The flights of Santos Dumont, in dirigibles or in airplanes, brought no decisive arguments for the military use of the "third dimension." Yet these flights had considerable impact on public opinion, and so contributed indirectly to the coming of military aviation. The flights of the Wright brothers, done with a catapult-type launching, showed the new perspectives open to aviation, and we will see further on the interest which the French General Staff manifested in these American builders.

A second question to which no satisfactory answer has been forthcoming is the following: Could Ader's AVION really fly? At the sight of this extraordinary apparatus -- modeled on the bat -- one is inclined to say no. Indeed the wing which moved in flight and the single wind tunnel test are enough to raise serious doubts.

It is surprising to note the degree to which this question of whether the AVION could really fly has monopolized the attention of those who interest themselves in this unusual man. Rarely has much attention been paid to his writings. This neglect may be traceable to the fact that in his major work, L'AVIATION MILITAIRE, he mixed technical considerations of questionable validity and prophetic views of the future. Beginning in 1897 he outlined what an air force should be, one which had its own air ministry and stood on equal footing with the other services. He described its organization, insisted on the necessity of supplying its most varied needs, proposed schools distinct from other military schools, and envisaged airports of various types, principal, auxiliary and mobile ones. He foresaw aircraft carriers and napalm; in the role of strategist weighing the power relationships in Europe he described with surprising realism a future Battle of Britain carried on by German air forces positioned in northern France. Many of Ader's ideas reappear in the works of General Billy Mitchell. It is very doubtful, however, that the American theorist could have had direct knowledge of the work of his predecessor. But Mitchell did have numerous contacts with French aviators. To what degree they may have been influenced by the man later called "the Father of Aviation" is a question difficult to answer. If General Roques went so far as to acknowledge Ader by deciding in November, 1911 that the Army's flying machines would thereafter be designated AVIONS, this does not imply that he adopted all of Clément Ader's innovative ideas.

When war came in 1914 it obliged aviators to draw from experience a body of doctrine which has a number of features in common with the theories developed in L'AVIATION MILITAIRE. To what degree were they inspired by those theories? Here there is a fruitful field for reflection and research. Indeed numerous comparisons might be made with foreign theorists, and particularly those who wrote before the war, such as Giulio Douhet and H.G. Wells.

CHARLES RENARD AND CHALAIS MEUDON

The numerous errors committed in the course of the war of 1870 led, as we have seen, to a considerable effort to improve national defense. A subcommittee on aerostation, attached to a committee on communications, came into existence; its secretary was a Captain of Engineers named Charles Renard. The subcommittee was originally housed in the attic of the Hotel des Invalides, but Renard prevailed upon Gambetta, who had considerable parliamentary influence, to get the subcommission and the aeronautical materials it had acquired after the war transferred to Chalais-Meudon, former birthplace of the Balloonists of the Republic. Funds were voted, laboratories set up, and in 1877 was created the Central Establishment for Military Aerostation, organizationally attached to the Corps of Engineers.

Once again appeared the individual initiative so important to the beginnings of aeronautics, and Charles Renard is a striking example of the phenomenon. Born in November, 1847, he entered the Ecole Polytechnique, graduated in 1868, and opted for a career in the engineers. During the War of 1870 he gave proof of his courage and leadership; he was a chevalier of the Legion of Honor at the age of twenty-three. Contrary to what has often been said, Charles Renard was not an unswerving devotee of the balloon. He believed that the future belonged to the heavier-than-air, but aware of the difficulties to be overcome, particularly in reducing the weight of powerplants in order to get satisfactory results, he believed that the balloon should first be exploited, then the dirigible developed, and finally the airplane.

Just to set up an aerostatic service proved a considerable task. With the aid of Lieutenant Arthur Krebs, despite limited resources, Renard determined the material needs in a few months and built the corresponding prototypes. At the end of 1879 the Ministry of War decided to create eleven aerostatic units: eight for campaigning (one per army) and three units for fortified sites. The balloons were rapidly built in series and a number of improvements made: a basket stabilized in vertical position, a steam winch utilizing two parallel capstans, and a continuously operating hydrogen generator mounted on a horse-drawn vehicle. Once it had proved itself on maneuvers, this materiél saw service overseas in the campaigns in Tonkin (1884), Madagascar (1895), China (1900) and Morocco (1909).

Having perfected the observation balloon, Renard turned to the study of dirigibles. Thanks to an eight-horsepower electric motor developed by Krebs, and chromium chloride batteries invented by Renard, the problems of propulsion were solved. On the 9th of August, 1884 the dirigible LA FRANCE took off from Chalais-Meudon with Krebs and Renard on board. It made a wide turn over Villacoublay and returned to its point of departure, having covered seven kilometers and six hundred meters in twenty-three minutes.

In 1898 the establishment at Chalais-Meudon became the Directorate of Military Aeronautics, with Renard as director. He continued his work on dirigibles, motors and propellers; at the same time he extended scientific knowledge in a broad domain. His work is little known. Conscious of the fact that his researches were in the interest of national defense, he did not publicize them; having something of an aversion to journalists, he declined the public exposure they might have given him. It would be interesting to determine to what degree the press, celebrating the accomplishments of Santos Dumont -- who DID like publicity -- may have diverted attention from the work of Renard, even among military authorities well placed to appreciate his accomplishments. One may gauge the importance of his work by examining his papers and the prototypes which are preserved at the Musée de l'Air. More than twenty motors driven by electricity, steam, hydrogen and gasoline, numerous instruments for measurement, propeller models, a whole theory of the helicopter propeller, all attest to the prodigious work of this man. His influence was great, both in France and abroad. It would be interesting to know just how far it reached. Samuel Langley and Igor Sikorsky knew his work. To what degree did they profit from it?

Renard also knew that it was necessary to continue work on the airplane. His studies on motors and propellers, though chiefly for application to dirigibles, could also be useful in heavier-than-air research. Aware of Captain Ferdinand Ferber's work in this direction, Renard succeeded in having him assigned to Chalais-Meudon in 1904. We will see later what he accomplished there.

Worn out by a life lived with great intensity, dedicated uniquely to work and research, Renard fell victim to depression and died on April 13, 1905 (he probably committed suicide). With him disappeared one of the greatest savants that France could boast of, a great patriot and a great soldier.

THE FIRST DIRIGIBLES

Despite all the experience amassed at Chalais-Meudon, the first dirigibles acquired by the French Army were not conceived there, but at the establishment of the Lebaudy brothers, Paul and Pierre. We will not go into detail concerning these acquisitions, but only give their names: the dirigible LE JAUNE, purchased in 1905, LA PATRIE, in 1906, then lost in 1907, LA VILLE LE PARIS in 1908; LA REPUBLIQUE, also bought in 1908, was destroyed in September, 1909 and its crew killed. This did not prevent the acquisition that same year of LA LIBERTE and COLONEL RENARD.

But from 1909 on the airplane appeared as a redoubtable competitor to the dirigible. At the 1910 maneuvers it showed its superiority. Yet the French Army, which still did not have great confidence in the airplane and watched carefully what was happening across the Rhine, did not abandon the dirigible, which was in principle reserved for long distance missions. By 1914 the French Army had received twenty-three dirigibles, furnished by an industry whose development would make a rewarding study for the historian. Of those twenty-three dirigibles, only six were operational when the war came. If the dirigible did not hinder the development of the airplane in France, it did lead her to abandon the captive balloon. In 1911 the companies of balloonists were once again dissolved, and once again on the grounds that they were not suited to a war of movement. Nevertheless balloons continued to be assigned to fortified places.

THE DEVELOPMENT OF MILITARY AVIATION

Captain Ferdinand Ferber, born in 1862, graduate of the Ecole Polytechnique and officer of artillery, was one of the great figures in military aviation. In 1898, while an instructor at the Artillery School of Application at Fontainebleau, he chanced upon old German journals which recounted the experiments of Otto Lilienthal. Ferber was particularly interested in the German's methods, so he built four gliders with which he attempted several flights. After tests with his fourth glider at Nice, the results convinced him that flight was possible. He continued his experiments and in 1901 he began a correspondence with Octave Chanute, through whom he knew the Wrights. Having built a biplane, he installed in it an eight horsepower Buchet motor; the craft was not so much an airplane as a motorized glider. Transferred to Chalais-Meudon at Renard's request in 1904, he built there a launching ramp of his own design and succeeded in May, 1905 in a controlled motorized glide of about a hundred meters. Unfortunately the death of Renard deprived him of valuable support. His machine was removed from the hangar on the orders of balloonists hostile to his work and destroyed in a storm. Ferber was obliged to leave the Army in order to continue his work at the plant of Léon Levavasseur. In 1908 Ferber had the satisfaction of flying in an airplane very similar to his craft that had been destroyed, and thus demonstrating what he might have accomplished two years earlier; on September 25, 1909 Ferber was killed in a banal accident while his craft was taxiing across the field.

A number of questions concerning Ferber's work need to be examined and resolved. Here again the scientific impact is undoubted, but the just value of his work has yet to be gauged. His quarrel with the balloonists and the indifference the high command showed towards his work likewise merit study. Then too his relationship with Chanute and the Wrights needs to be better understood, in order to better grasp the reciprocal influences among the French and American pioneers at the very dawn of the century.

CHALAIS-MEUDON AND AVIATION

The disappearance from the scene of Renard and Ferber did not halt the valuable work on aviation being done at Chalais-Meudon. We cannot treat this activity in detail, but we should note the achievements of Captain Jacques Saconney concerning military kites and the work of Captain Albert Etévé on speed indicators and on stabilizers destined to be used on the Wright airplanes. We should also mention Captain Dorand, who would become in 1906 head of the technical services for aeronautics; from 1906 to 1914 Dorand carried out numerous studies of helicopters and airplanes. Despite the importance of these accomplishments, they did not convince the General Staff of the value of the airplane. That was done by two notable exploits in 1908 and 1909: the first closed-circuit kilometer, flown by Henri Farman, and the flight across the Channel by Louis Blériot.

HENRI FARMAN'S CLOSED CIRCUIT KILOMETER

On January 13, 1908 Henri Farman first succeeded in flying an airplane in a closed circuit of one kilometer. To be sure, this event may seem of no great importance, coming as it did some four years after the flights of the Wright brothers. But in fact the Wrights had not succeeded in taking off in the strict sense. They used a launching system which propelled them into the air. Until Farman, all attempts to take off by the sole power of the motor had been carried out with the joystick pulled all the way back, which resulted in the airplane making nothing by grasshopper jumps. Farman discovered, simply by trying it, that in order to take off one should first move the stick forward, which enabled the craft to gain speed. This discovery made his feat possible, and it also enabled many other aircraft to take off for the first time. This historic flight gave fresh incentive to French airplane builders, but what was more important for the future of military aviation was Blériot's flight across the Channel on July 25, 1909. The possibilities of the airplane had been proved. Those possibilities were confirmed at the international meeting at Reims-Béthény in August of the following year. There the French excelled, but also a young American aviator named Glenn Curtiss attracted attention.

The French Army had been considering the purchase of airplanes since 1906, when there had been contacts with the Wrights. But the parties could not agree on contract terms, so the negotiations were dropped. But the Wright brothers had not given up the idea of selling their planes in Europe and in 1908 they made a series of demonstration flights in France in which they beat all the existing records and proved to the French public the reality of what they had accomplished in the United States.

But the Wrights' designs were not the answer for the future, despite the improvement of their flight characteristics through the use of Etévé stabilizer. At the Reims meeting they were outclassed by French aircraft. Even so, in September, 1909, the Army bought two Wright biplanes, as well as two Farman biplanes and a Bleriot monoplane. All of these planes were for the Corps of Engineers. The Artillery protested, received funds of its own, and acquired seven aircraft (3 Farmans, 2 Wrights, and 2 Antoinettes). At about the same time ten officers received their civilian pilot's licenses, seven engineers and three artillerymen.

From the very beginning of military aviation there was thus a rivalry of sorts between artillery and engineers. The Corps of Engineers, stressing its long association with aeronautics, felt it should control airplanes, which were a natural complement to the balloons it already possessed. The artillerymen emphasized their special needs for directing artillery fire and their competence in motors, since they maintained the Army's fleet of motor vehicles. The dispute likewise covered the employment of airplanes. In any event it was prejudicial to the development of the new arm, so the high command sought to resolve the quarrel by deciding in September, 1907, to place both the aerostatic and the aviation services under a single authority: the Permanent Inspection of Military Aeronautics.

General Roques became the first Inspector of Military Aeronautics. The Artillery-Engineers rivalry continued nonetheless. The two establishments at Vincennes and Chalais-Meudon continued working, each according to its own lights. Despite the obvious lack of cooperation, aviation did not suffer much from this. There is even some reason to believe that the competition served to quicken activity rather than slowing it, leading to a good deal of reflection on how the airplane could be used in war.

In 1910 airplanes participated in the general maneuvers in Picardy, which were rich in lessons for the future arm, even though it was on that occasion that General Ferdinand Foch, future generalissimo of the Allied armies, is supposed to have made his celebrated judgment on the airplane: "For sport, yes; for war, zero." The year 1910 likewise saw the creation of two schools of military aviation at Chalons and Vincennes. Finally, in November, 1910, it was decided to hold a military aircraft competition to supply the Army with craft suited to military missions.

The competition was held in 1911 and revealed the good qualities of Nieuport, Breguet and Deperdussin aircraft all fitted with rotary engines; but the military could not immediately profit from those qualities for lack of funds. The same year airplanes began to be used in the colonies, a subject whose whole history has yet to be written. And other important events occurred in this period, such as the creation of the military pilots' license, more comprehensive than the civilian license, and the creation of specialties such as aerial observers and mechanics. Elsewhere the Italians in Libya used Blériot planes purchased in France for the first wartime reconnaissance flights.

As military aviation developed, so did its civil counterpart, and the aeronautical industry began little by little to grow beyond the purely artisanal stage, producing in the year 1911 some 1300 aircraft, 1400 motors, and 8000 propellers. Production increased further in 1912: 1425 planes, 2217 motors, and 8000 propellers.

The year 1912 was to be of considerable importance to aeronautics because in that year the National Assembly voted the Law of March 29, organizing military aeronautics. Flying personnel were given a distinctive status and a definite organizational framework. Other changes were decided by executive decision, such as the adoption of the tricolor roundel, the creation of a distinctive uniform, flight pay, etc. Then at Longchamp on the 14th of July, 1912, on the occasion of the revue of troops, M. Armand Fallières, President of the Republic, presented the aeronautical service with its own flag. Yet there were still problems under the surface, and they could be seen in the fall maneuvers, where a number of points were raised, and where the engineers particularly raised the question of applying armor to airplanes.

The country as a whole began to be aware of the new arm. The patriotic enthusiasm which welled up took the form of a national fundraising campaign launched by the newspaper LE MATIN, which amassed some 4,000,000 gold francs for the purchase of military craft. Yet in spite of this a governmental decision of April 16, 1913, stripped the Permanent Inspection of Aeronautics of part of its functions, placing the flying units under the direct authority of the army corps commanders in whose territory they were stationed. This was the beginning of a vast dispute which would take a long time to die down -- a dispute over the degree of autonomy the new arm should have, given its unique characteristics and methods of operation.

In 1913 the decision had no immediate consequences, but that was not at all the case with another change which marked the victory of the Artillery over the Engineers, symbolized by the replacement of General Hirschauer (who had himself replaced General Roques as Permanent Inspector of Military Aeronautics). The new Director was an artilleryman, General Jean-Baptiste Bernard, whose most salient characteristic was his total ignorance of the problems of aeronautics. This appointment was a serious matter, for war was approaching, and the first act of the new Inspector was to order a change which satisfied the artillerymen but had an adverse effect on aircraft production. General Bernard ordered that henceforth airplanes be armored; given the feeble horsepower of the motors of the day this decision led to the production of planes which were slow, heavy, and unsuited for operations. At the beginning of 1914 there was another reorganization. The Permanent Inspection was abolished in February and the Directorate of Aeronautics created in April, with General Bernard at its head.

Through all of these changes aviation continued to progress. The training of pilots had been accelerated. By the outbreak of the war a total of 1250 aircraft had been purchased, enabling the Army to have a front line force of 126 at the start of hostilities. There were also reserve stocks sufficient for the brief war which the Army planned to fight -- so brief in fact that no consideration was given to supplying additional pilots or aircraft. Upon mobilization General Bernard had the schools closed and sent aircraft factory workers to the front. These decisions would all be reversed as soon as the premise that inspired them showed itself to be false, but not before some damage had been done.

The significant development of aircraft we have seen in the years between 1909 and 1914 was not accompanied by the formulation of any brilliant ideas on their use. For the high command the only conceivable role for aviation was that of scout or observer for the ground troops. All of the experiments carried out to better determine the possibilities of the airplane were unknown at the high, decisionmaking levels. Yet the Michelin Cup for the best performances in aerial bombing had revealed the airplane's possibilities in that domain. Likewise, the trials of armament and firing from planes had given promising results. Photography and wireless had been tried successfully. The high command remained skeptical. When Captain Jacques Faure submitted a report on the use of the 37mm cannon in the Voisin aircraft, it was judged "very interesting work which reveals in the officer who carried it out a spirit of inquiry which ought to be encouraged, but the trials themselves have more in common with Jules Verne than with reality."

In the period 1870-1914 the development of aeronautics began slowly in France, then moved rapidly after 1909. The period is a valuable one to study because it offers a whole range of viewpoints and attitudes which would reappear throughout the history of military aviation both in France and abroad.

The first characteristic of this development is that it owes nearly everything to individual initiative, rather than to decisions of the high command. It depended upon men like Clément Ader, whose ideas had for a time the good fortune to be sustained by an extraordinary statesman who was himself a savant and an innovator -- Freycinet. But achievement here was cut short by the decision of a general serving as minister of war. There was also the work of Renard; he was sustained by the ardent patriot Léon Gambetta, one of the great figures in the political history of France; but much of Renard's work was also virtually unknown in the higher circles of the Ministry of War. Then there was Ferber, paying out of his pocket the mechanic who helped him build his plane at Chalais-Meudon; Ferber too would know many disappointments and would be obliged to sacrifice his career in order to develop an aviation which would prove essential to the defense of his country. All of these men were innovators driven by an unshakeable faith in the future of the airplane, a faith which led them to pursue research which was still largely ignored when the war broke out.

Yet the work of such men was not without its effect. Little by little, aeronautics progressed in France. Ferber saw clearly when he said that his work was not so much that of an inventor as that of a propagandist. This points up a curious contrast -- and one worth studying -- between a public enthusiastic over aeronautics and a General Staff which was silently skeptical. The facts are indisputable. The exceptional development of the aeronautical industry can be measured. Its success in sales abroad was manifest, the leadership of France in this domain was evident throughout the world. Despite all of this, official circles remained prudently uncommitted. This attitude can be explained and indeed justified. Appropriations for the French Army were limited. It would not be wise to sink them into adventurous schemes. War was approaching and there was no certainty that the airplane would play a major role in it.

Are these arguments the genuine and the sole reasons for the action of the General Staff? Were there not other considerations more doubtful in their sincerity and other reasons less noble which worked against the emergence of the airplane? And did the quarrels between partisans and adversaries of the heavier-than-air and the antagonisms between engineers and artillerymen also have an inhibiting effect?

Even if aviation had not been able to convince ruling military circles of its effectiveness, it made real progress. With the exception of Germany after 1912, no other country could boast of such activity. Here again some very useful comparisons might be made concerning general attitudes and those characteristic of certain countries. England and Russia, for example, though as deeply concerned about war as France was, made no effort in the air comparable to hers. What factors operated here? Did they also operate in the case of the United States? To be sure, for the latter country, the nature of the problem is different. The major principle of American foreign policy was still isolationism, even though the war with Spain and the campaign in the Philippines had done some violence to that principle. Though isolationism still offered valid arguments, there may have been other considerations which militated against the development of American military aviation, and indeed subsequent actions on the part of American military leaders suggest that was the case. But here again it would be interesting to make a comparative analysis of the American and European experiences. It would also be worthwhile to know how the American military viewed the innovations being made in France. Foreign military attachés followed French maneuvers closely, and particularly those of 1912, which were valuable in the lessons they taught. No doubt the American military attaché made reports. Could they be found? Then too the Wrights and Curtiss were in frequent contact with French aviators, who probably made judgements in turn. Have these been taken into consideration by historians?

The war which came in 1914 would both continue and modify pre-existing trends in military aeronautics. Aviation would be called upon to play a major role, along the lines predicted by air strategists of the preceding years. All of the errors committed on the eve of the conflict were not mortal ones; their consequences, which might have been serious, were soon overcome.

THE GREAT WAR 1914-1918

When war broke out in 1914 the air forces of the great powers were still very small. Germany's effort in the air, though late in its inception, had borne fruit: 232 airplanes and 12 dirigibles, as well as a very large number of Drachen balloons. Austria-Hungary possessed 56 planes, while France had 134 planes and 6 dirigibles. Great Britain went to war with 84 airplanes and 4 dirigibles, and Russia with 190 airplanes.

The balance of forces -- 288 airplanes and 12 dirigibles for the Central Powers and 408 planes and 10 dirigibles for the Entente -- had in itself no particular significance. The aircraft of 1914 would not struggle for command of the air, since the general staffs of the belligerents all planned to use them exclusively for aerial observation on behalf of the ground forces.

With mobilization France was able to increase very rapidly the number of escadrilles available; they passed from twenty-three at the beginning of August to twenty-nine at the end of that same month. The total personnel thus committed to the air units and their ground support was 14,000, including some 250 officers and about the same number of non-commissioned officers. If we compare this figure with the vast numbers of effectives in the French armies, we can see the very modest role assigned the aeronautics at the outset of the war.

THE ESCADRILLE

The basic unit for French aviation, from the beginning of the conflict to its end, was the escadrille. To make effective use of archives and other documentation for this period, the researcher must understand what these small units were and how they were designated. At the beginning of the war each escadrille had six aircraft, save the cavalry escadrilles, which only possessed four. The number of officers and men in an escadrille was about fifty. By the end of the war escadrilles on the Western Front had from twelve to fifteen planes and complements of from 100 to 150. These units were numbered beginning with "1," in the order in which they were created, though, to be sure, there were some exceptions, as well as some gaps in the numbering system. The number was preceded by the initials of the particular craft used by the escadrille, and while the complete list of these initials cannot be given here, we might note some of the more important: MF for Maurice

Farman, HF for Henri Farman, V for Voisin, VB for Voisin bomber, N for Nieuport, BR for Breguet, and SPA for Spad. While the escadrille kept its number, it might change initials as it changed aircraft. For example George Guynemer's escadrille was known successively as MF3, BL3, N3, and finally SPA 3. These escadrilles of the Great War, lineal ancestors of units currently in the Armée de l'Air, are generally designated by historians by the last appellation they carried in that war.

AEROSTATS, 1914-1918

Dirigibles proved disappointing when used on the land fronts but much more valuable on the seas. In 1914 the Germans had already envisioned long-distance reconnaissance and bombing roles for their Zeppelins, but very soon the great vulnerability of those aerial mastodons led the German authorities to use them only under cover of night, save on the Eastern Front, where they flew by day as late as 1917. The Zeppelins made terror raids on large cities, and while the results obtained over Paris and London were very modest, those raids, and others against England by Gotha bombers, had considerable indirect effects. In Britain they led to a clamor for an independent Royal Air Force and the use of aircraft carriers, and in France they led to the formation of night fighter units for the defense of Paris -- units which were singularly ineffective, but which would make an interesting study nonetheless.

The Germans used seventy-eight dirigibles for long distance bombing and reconnaissance, but a goodly number of those fell victim to British defenses or to bad weather conditions -- conditions difficult for the Germans to predict since weather fronts generally moved from west to east. Seventeen German dirigibles were shot down, six were destroyed on the ground by the British, and thirty were lost to bad weather or through breakdowns.

The Allies used the dirigible extensively for convoy, escort and antisubmarine warfare. The British used no fewer than 200 airships for this purpose, and the French Navy employed about forty. This is an interesting aspect of the war, still largely unstudied. It would be useful to know how deeply the lessons learned about using dirigibles against strong antiaircraft defenses or in bad weather really sank home, and why in the face of those lessons air power spokesmen continued to advocate the airship for long range bombardment. Was it the aim of such spokesmen, Mitchell among them, simply to defend a weapon which seemed unsuited for such use, or were they trying to convince public opinion of the value of strategic bombing itself? Certain it is that in the early twenties the dirigible alone possessed the range and load-bearing capacity for threat and reprisal at the strategic level.

From the very beginning of the war the German Drachen observation balloons had proved their usefulness, so the French high command set about reconstituting the units it had earlier abolished. This was done rapidly, both by searching depots for pre-existing material and by developing new equipment. After borrowing the design of the German Drachen and improving upon it, the French Army adopted a remarkable balloon devised by Albert Caquot. It was subsequently standardized by all the Allies and by the enemy as well after the wind currents carried several Caquot balloons over the German lines. This was the basic design used by all the aerostatic services, and as barrage balloons, during both world wars.

Aerostation was of considerable importance during the war of 1914-1918, though the exploits in military aviation have tended to obscure its role. In truth balloons and airplanes did not compete so much as they complemented each other. Airplanes could cover the whole of a zone and take photographs from overhead which were more accurate than the oblique photographs the balloons produced; then too the airplane could take photographs deeper in enemy territory. On the other hand the balloon offered a permanent vantage point and the only means for a constant surveillance of the battlefield, instantaneous fixing of the position of enemy batteries, and the direction of friendly artillery fire on objectives at close or medium range. Little by little balloons began to appear all along the front, attached to armies or army corps. Though very valuable in positional warfare, they also proved sufficiently mobile to follow troops in movement when the Allied offensives began in July, 1918. This time the old arguments which had sent the balloons to storage depots did not prevail and the aerostation arm of the service proved its usefulness by the end of the war. It had also found the full range of possibilities and the limitations in that usefulness. It was very clear that the use of balloons along the front lines required air superiority, as was shown at Verdun. Then too the balloon's vulnerability to enemy fighters had become greater as the fighter's speed increased, giving it greater chance of surprise. But these considerations were soon lost sight of, for in the Ecole Supérieure de Guerre the employment of balloons came to be treated almost as if the enemy fighter force did not exist.

THE AIRPLANES

If the role which fell to the aerostats changed little in the course of the war, this was not at all the case with airplanes. At the outset the only prescribed function for these was observation on behalf of the armies. Even here there was initially little confidence in the information which came from aerial observation. When Corporal Louis Breguet and his observer Watteau reported a change in the direction of march of Von Kluck's army, French military intelligence, which already knew the routes prescribed for the German armies, paid no attention to the report. It paid no more attention when airplanes under the orders of Captain Bellenger, chief of aviation for General Manoury's army, confirmed this development. It was only thanks to the openmindedness of General Joseph-Simon Galliéni, and confirmation supplied by a patrol of dragoons, that the French high command accepted and launched the victorious counteroffensive along the Marne.

The results obtained by aviation soon convinced Army authorities of the value of the new arm, though the endorsement was far from total, and what improvements came in materiel and techniques came more often from individual initiative than through directives from above. At the beginning of the war this could be seen in a number of ways, and here we may cite a few of them. The use of cameras aloft was not a new idea. Aerial photography had already known some successes, particularly with pictures which Lieutenant Etévé took from a captive balloon in 1907 during the siege of Casablanca. Experiments aboard aircraft had likewise given good results. Yet the first aerial photographs taken in the war were by observers using their personal cameras. Fortunately this was so successful that the Army's hierarchy was convinced and aerial photography was functioning formally before the end of 1914.

Something of the same evolution occurred with fighters. Firing experiments from aircraft had been done before the war, but no-one in authority put much faith in the prospect. It was the aviators themselves who sought out the best means to strike at enemy craft which they encountered in the air. Some carried rifles, while others jury-rigged machine guns in their planes. So it was that Sergeant Pilot Joseph Franz and his mechanic Quenault, flying a Voisin, shot down an Aviatik on October 5, 1914. This first air victory was to be followed by many others, and by a rapid evolution of aerial armament.

One new technique, that of firing through one's own propeller arc, was the result of individual initiative. Before the war studies had been made, both in France and abroad, for the use of synchronized fire through the field of a rotating propeller, but this work had subsequently been forgotten. Roland Garros, who had worked with the engineer Saulnier, got the commander of his escadrille to send him to the rear to take up this work again. At the time France was using the Hotchkiss machine gun, whose mechanism did not lend itself to synchronization, so Garros had to settle for a bullet deflector which he designed. Back at the front, with his parasol Type M Morane-Saulnier he proved the value of his invention by downing three German planes in seventeen days. Unfortunately he had to land in enemy-held territory and could not destroy his plane. The Germans studied his deflector and asked Fokker to develop a similar device. But Fokker went one better and developed a synchronizing device. Thus the fighter arm was born of individual initiative, followed by official sanction and encouragement.

The last example of the role of individual initiative concerns bombing. We have already mentioned the bombing competition sponsored by the Michelin brothers, which did not convince French officialdom. But pilots flying over enemy objectives thought naturally of some means to destroy them. French aviators were all the more inclined to such thoughts since it was the Germans who first dropped bombs on Lunéville on August 3, 1914. They repeated this technique several times during the month, while Lieutenant Césari and then Lieutenant Prudhommeaux, attacked the dirigible shed at Metz-Frescaty on August 14 and 15, damaging a Zeppelin. Here too, makeshift was the basis for these first experiments -- makeshift and a spirit of enterprise as well, since the first bombs were devised from shells obtained from the artillery. Here the French high command soon made bombing official, creating on September 27, 1914 the first bombing GROUPE, composed of three escadrilles equipped with Voisin pusher planes.

THE REVITALIZATION OF FRENCH AVIATION

The experience of the first engagements in the war soon demonstrated that the conflict would be a long one and that aviation would play a significant role in it. Consequently aviation would have to be developed further, more planes would have to be built and more pilots trained. Major

Joseph Barès, head of aviation, was hard at work, and on October 8, 1914, the high command put forth a plan to increase air strength from 31 escadrilles to 65 in the space of three months. To put on the line the 384 planes this plan called for, it would be necessary to produce 2400 aircraft and 3400 motors. It was also necessary to reopen the schools, which had been closed at the outbreak of war on orders from General Bernard. That officer, manifestly incompetent, gave up his command on October 25, to be succeeded by General Hirschauer, who had preceded Bernard as permanent inspector of aeronautics. Hirschauer launched the development plan, but its objectives were too ambitious. In March, 1915, there were only 53 escadrilles at the front. The great majority of these units were doing observation work and artillery spotting for the various army corps. Some had been collected into groups specialized in day bombardment; still others, towards the end of 1915, would emerge as the fighter arm.

ARMY CORPS AVIATION

We have seen that at the beginning of the war the army staffs had only a limited confidence in observation work by aircraft. This hesitation tended to disappear as the airplane proved itself, though as late as September, 1915, the high command refused to believe aviators' reports that a certain trench, the "Trench of the Tents," was still held by French troops. It took three days and the arrival of an officer from the trench in question before headquarters was convinced and ordered a halt to the bombardment which it was directing on its own troops. But subsequently observation made by airplane came to be regarded as the most trustworthy. These observations were reinforced by aerial photographs, whose usage spread; soon there was better equipment, as well as photographic sections which could rapidly develop film brought in by aerial observers. Before long the equipment included cameras for vertical, oblique or panoramic work. At the same time interpreters emerged to draw the maximum amount of information from the photos they analyzed.

Army corps aviation also helped regulate artillery fire. At first the aviator noted the points of impact and then landed near the battery to relay his observations. This was a tedious process which slowed bombardment, so by 1915 on-board wireless sets appeared so that the observer could relay his information by Morse code to the battery, which signalled him by using signs. This technique was highly successful and remained in use to the end of the war.

Army corps aviation grew as the ground units came to realize its value. Beginning with thirty-odd escadrilles at the end of 1914, the service grew to 50 escadrilles in October, 1915, then 80 in the spring of 1916, and finally 126 at the end of the war. At the outset the planes could carry out their missions without serious threat from the enemy. Fighter planes scarcely existed, so that the only danger came from rifle fire of the enemy infantry. But from mid-1915, the enemy's air attacks became more frequent and more dangerous. Soon it was clear that the observation planes used at the beginning of the war lacked the performance to carry out their missions.

To remedy this situation two different measures were taken, first to improve the performance of army corps aircraft, and second, to provide them with fighter protection. The first measure meant the recall of pusher type aircraft, highly vulnerable to attacks from the rear, and their replacement by traction machines, protected by the gun mounting system devised by Etévé in mid-1915. A new generation of planes would not appear until 1917 so that army corps aviation had to endure all of 1916 with older models.

The other solution was to protect army corps aircraft with fighters. There was general agreement on this solution, but not on how to bring it about. Some believed -- and as we will see, the example of Verdun seemed to bear them out -- the best procedure was simply to sweep the enemy aircraft from the sky in a specified sector, thus affording indirect protection to the observation planes. For others, this distant and indirect protection did not suffice, and a close accompaniment seemed the answer. This was the beginning of an argument that would long continue between observation aviators who wanted a guard dog at their sides, and fighter pilots who felt, often with some justice, that flying close escort to planes of inferior performance hampered them in their contests with enemy fighters.

In truth this quarrel, which began long before Verdun, proved the fighter pilots to be right. In the first part of 1917 there were heavy losses of observation planes in connection with the Nivelle offensive, but by the autumn of that year Allied air superiority was beginning to manifest itself and several remarkable new observation craft appeared: the Caudron G. 6 twin-engine three seater, the Salmson 2A2 and above all the Breguet XIV A2.

BOMBARDMENT AVIATION

We have seen that aerial bombing developed very early, and in the French service led to the formation of the first bombing group, GB1. The planes chosen were Voisins and Farmans, with pusher propellers and a low speed. Fortunately, into 1915 neither antiaircraft fire nor enemy fighters were much in evidence, so that the greatest risk was from mechanical failures. Under the guidance of two exceptional leaders, Major de Goÿs and Captain Happe, the bombing units evolved both in materiel and in technique. In the beginning the bombs were simply thrown over the side, the aiming being done with the aid of nails driven into the fuselage. Soon racks were devised for the bombs, which were often 90 mm or 155 mm artillery shells with fins added. Work began on a true bombsight, and at the same time bombing as an act of individual initiative gave way to mass bombardment, facilitated by the creation of such units as GB1.

Major de Goÿs pushed the training of his crews and on May 27, 1915, he led an eighteen-plane bombing raid against the Yperite and explosives plants of the Badische Analin Company at Ludwigshafen. The raid was a success, although engine failure forced de Goÿs down behind German lines and he was taken prisoner.

For his part Captain Happe, commanding Escadrille MF25, equipped with Farman XIs, had the idea of flying in V formation in order to get a better concentration of bombs on target and also provide mutual defense against attacking fighters. It is significant that this formation developed by Happe had been foreseen some fifteen years before by Clément Ader.

Though three bombardment groups had been created (GB2 and GB3 on September 15, 1915), the high command decided because of heavy losses, to abandon strategic bombardment and to concentrate bombing efforts on targets immediately behind the battlefield. The hopes earlier placed in aviation had led the high command to foresee in July, 1915 the eventual creation of 50 escadrilles of 10 planes each. After December bombing units were switched to night missions and the projected number of escadrilles reduced to 31.

Night bombardment as it was practiced in 1916 and 1917 presented numerous difficulties. Navigation was very demanding, location of targets difficult, and the return to the home field uncertain, despite the lights set up to guide the navigators. In truth, given the technical limitations of the period, bombardment of this type was feasible only on bright, moonlit nights. The number of attacks possible was thus limited, as well as their effectiveness. Even so, bombing units took part in all the land battles of 1916; moreover a few raids which could properly be called strategic were carried out by small units or single planes. Among these raids we might mention that carried out against Essen on September 24 by two Sopwiths, and that of Captain Beauchamp, who took off from Luxeuil in a specially equipped Sopwith and landed in Italy eight hours later, having bombed Munich on the way. There were also some daylight raids, including one carried out in reprisal for German raids on French cities: Escadrille C66, led by Captain de Kérillis, bombed Karlsruhe, producing eight hundred casualties there, while a third of the raiding planes fell to German fighters.

The year 1917 was in many ways similar to 1916; it saw many night raids on the enemy's rear, against railroads, camps, troop concentrations, etc. From about mid-August on there was the "airfield war," as it was called, as bombers on both sides concentrated on these installations. Then French bombers carried out night raids on industrial centers in the eastern part of the country which was under German occupation. The choice of targets eventually stirred much controversy, for the factories of the industrial magnate M. de Wendel were never bombed. It would be interesting to compare this controversy with those which developed after World War II, notably over the bombing of oil targets.

It was only in 1918 that extensive day bombing reappeared, thanks to the massive employment of the Breguet XIV B2, which began to appear in the formations in the autumn of 1917. To strengthen the power of this bombing fleet, it was grouped into two ESCADRES, each of three groups, with each group containing three fifteen-plane escadrilles. From July, 1918 on the bombers were accompanied by twin-engined Caudron R. XIs which served to protect the flanks of the formations. These Caudron planes were the origin of a double controversy: the first over the question of the best two-seater or "multiplace" fighter; the second over the so-called BCR or multipurpose aircraft concept. The new bombing escadres were joined with an escadre of fighters to form a GROUPEMENT, as we will see later.

While the return of day bombing involved strikes against targets close to the battlefield, the high command foresaw a return to strategic bombing of enemy industry, but the Farman 50, designed for such raids, did not appear in time to be used with any effect. Alongside the day bombing unit two GROUPEMENTS of night bombers carried out massed raids to the end of the war.

It is easy to understand why the French high command should have abandoned strategic daylight bombing to switch to night operations, then back to daylight raids when suitable planes became available. At the same time the Germans and the British were developing strategic bombing forces, the Germans used them essentially in terror raids on London and Paris, while the British bombers flew retaliation raids. It is also easy to understand the reasons which impelled the British to adopt a bombing force, independent of land operations -- Britain's insularity explains this choice perfectly. But it is harder to understand why a continental power like Germany should do so, wasting her air strength in terroristic raids on great cities, when that strength might have been better used in the great land and air struggle along the Western Front.

The French position was certainly justified and conformed realistically to the technical possibilities of the era. In 1918 even the Handley Page V/1500, which came too late to be used, would not have enabled the British to solve the problems of a strategic air offensive, namely how to transport far from one's bases a sufficiently high tonnage of bombs to have some effect, and then deliver them over a heavily defended target. Navigation at night or in bad weather with a safe return to base presented problems that would only be solved much later. But the idea was worth pursuing. Only the British, and later the Americans, would follow through with effective strategic bombing forces in World War II. Despite the desires of many of her aviators, France would not pursue this course, but would give in to the wishes of the land army and its leaders. As for the Germans, preoccupied with the necessity of developing an effective tactical air force, they were compelled to improvise in their air offensive against England in 1940. This they were able to do because the fall of France offered them bases very close to their enemy. But later, against Russia they would feel the lack of a long-range bomber which could strike Soviet industrial targets.

It is clear that there were a number of choices in the matter of bombing, and the debate over these options would arise in many countries, often pitting the air service against the other branches of the armed forces. The lessons of 1914-1918 were various and sometimes contradictory. What seemed proven was the effectiveness of bombing the rear areas of the enemy army, and as we shall see, the possibilities for direct intervention on the battlefield.

THE FIGHTER ARM

If the fighter was born in the spring of 1915 with the first successes in firing through the propeller arc, the fighter arm was not organized until later. Indeed up to the battle of Verdun in 1916 fighters were only used in sporadic fashion by both the Germans and the French. In the Fokker E-1 equipped with a synchronized machine-gun the Germans possessed a formidable weapon, but for fear that weapon might fall into Allied hands, they forbade flights "over the line." Then too the small numbers of these aircraft, a hundred scattered along a front 600 kilometers long, kept them from being a mortal danger to French army corps aviation. Then in February, 1916, everything changed. Germany decided to deliver a mortal blow to France at Verdun. This blow was to be preceded by an artillery bombardment of unprecedented ferocity. In order that the bombardment and the infantry attack following it might have the greatest possible effect, it was necessary to "blind" the French artillery, reducing both the effect of their counterbattery fire and of that on the German infantry. To do this the Germans had to bar from the skies both French observation balloons and army corps planes. For this purpose the Germans gathered around Verdun a mass of 180 airplanes which were to obtain command of the air in the Verdun sector. At the same time carefully directed fire at French balloons and their winches would keep the "sausages" out of the air.

The initial success was total. On February 21, 1916 German aircraft took control of the skies and effectively barred French aerial observation when the French Army needed it most. In this crisis Joffre ordered Major Tricornot de Rose to regain air superiority. Calling on all of the best pilots of the day, de Rose gathered in the Bar-le-Duc sector fifteen escadrilles equipped with Moranes and Nieuport XIs, the latter of which had just appeared. In these escadrilles were men like Guynemer, Charles Nungesser, Albert Deullin, Raoul Lufbery, and above all, Jean Navarre, "the Sentinel of Verdun." Operating en masse, these escadrilles retook control of the skies and made it possible for French observation planes and balloons to return to their vital work. From April on the air situation had stabilized and this was followed shortly by the stopping of the German offensive. Verdun is rightly considered the hardest battle of the Great War. It has come to symbolize the suffering and the determination of the foot soldiers. It is also noteworthy, though little noted, that Verdun was the first great victory for French wings.

If Verdun saw the first real organization of the French fighter force, it also saw the birth of a principle which lay at the base of air doctrine: obtaining local air superiority as a precondition to opening a land offensive. This design was adopted by the high command and applied in the battle of the Somme, which began in July, 1916. During the battle our fighters took and held control of the air. They developed their tactics and confirmed their earlier achievements. But air superiority depended upon a number of factors still not perfectly understood. This could be seen during the Nivelle offensive in the spring of 1917, when French fighters adopted an offensive posture which led them to carry out raids over enemy-held territory. These were often strokes delivered in a void. German fighters often refused battle there; while French fighters were at great distances from their bases, the Germans profited from their absence to penetrate over French lines and attack army corps aircraft. This faulty doctrine, though not entirely false, was the basis for a new conception, opposed to that adopted after Verdun.

In truth air superiority, even on a restricted front, is never more than momentary. Massing fighters in order to achieve that superiority exposes unprotected observation planes to certain destruction. So it seemed better to give up this practice and to use fighters for close support of the observation planes indispensable for land operations. To get the best results it seemed logical to place the fighter escadrilles under direct orders of the army corps commanders. This theory, adhered to by the land army for many years, even after the Second World War, has been attributed to the heavy losses in observation planes suffered along the Aisne (Several French and British authors have presented this version of the story).

The truth is far different. It is true that the British attacks in Artois cost them very heavily in "Bloody April" -- 139 planes. It is also true that German aviation dominated British in this period, leading British admirers of the Richthofen Circus claim German air superiority on the entire front. The idea of Allied air inferiority was also a convenient means for French Army leaders to explain their failure at the Chemin des Dames. But this argument does not stand up under close examination. From April 6 to 25 French losses were 33 planes (16 observation and 17 fighting craft). In the same period the Germans lost 6 observation planes and 63 fighter planes. Two things are clear: even if French fighters might have given better protection to observation planes, the losses among the latter were not catastrophic; then too, French fighters obviously had the upper hand over their German opponents. It would be interesting to know why these easily available figures were not used in debating the various doctrines, and also why the myth of German air superiority over the Aisne has been so tenacious of life, particularly in recent times when there is no propaganda value in denying the truth. We should note in closing this section that French fighters went on to obtain solid successes in the limited offensives which General Pétain directed from August to October.

COMBAT AVIATION AND THE AIR DIVISION

On March 2, 1918 appeared the first operations manual for combat aviation entitled NOTES ON THE USE OF COMBAT ESCADRES AND GROUPES IN BATTLE. Fighter units were put in combat GROUPES assigned to armies for protection of observation planes, and into combat ESCADRES under the orders of the high command. The prescribed role for these combat ESCADRES was offensive operations -- seeking out and destroying enemy planes and Drachen balloons attacking troops on the ground. An ESCADRE was made up of three GROUPES each of three escadrilles; an escadrille contained fifteen planes and 150 men. To assure its mobility, the ESCADRE had ground transport and support facilities. At the same time day bombardment ESCADRES were organized, each containing three GROUPES of three escadrilles each; the escadrille was equipped with fifteen Breguet XIV B2 bombers.

In order to increase the effectiveness of these ESCADRES, the high command decided to join an ESCADRE of fighters to an ESCADRE of bombers, to form a GROUPEMENT. The two GROUPEMENTS created in March, 1918 carried the names of their commanders: Groupement Féquant and Groupement Ménard. Thanks to this organization, French air units "reserved" for use by the high command became independent, not of the fighting on the ground (that would have been heresy), but of ground organization. The high command could employ this aviation over the entire front, and it alone had the breadth of vision to use it where it would yield the greatest benefit. This could be seen at the time of the German offensive in Picardy, where French fighters eliminated their opponents in two days, opening the way for the bombers to intervene in the land battle.

The results were rewarding, but to enhance them even more, the high command decided to carry further the centralization of command. On May 14 it created an air division, sometimes known as the "Duval Division," after its commander. It contained the two day-bomber GROUPEMENTS already mentioned, plus two night-bombing units, GROUPEMENT Chabert and GROUPMENT Villomé. This air division, whose GROUPEMENTS would be rechristened brigades, made a considerable contribution in attacking enemy troop concentrations. Results were very rewarding, and a number of examples could be cited, for example the neutralization of a heavy German concentration at Villers-Cotterêts by 120 bombers dropping 7000 bombs on the single day of June 4, 1918. But the division's finest hour came in July, when it was thrown against German troops driving for the Marne, since there were no French ground forces immediately available for this task. By air attacks alone the breach was closed and the German drive across the Marne brought to a halt. German troops would later speak of those days, during which they were subjected to incessant aerial attacks, as a veritable hell. Here we find ample demonstration of two principles: first, that the commitment of aviation should be done at a fairly high level in order to concentrate its effects; second, that direct engagement of air strength into the land battle could have decisive results. The Allied offensives of 1918 would use masses of tanks supported by masses of planes, demonstrating the value of the tank-airplane team once air superiority had been acquired.

Why is it that in the inter-war period people remembered only the 1915-1917 phase of the war? This question is seldom asked and the answer never completely satisfying. Why did everyone accept the idea that a continuous front could not be pierced? And why was aviation consigned to its servile role in support of army observation, despite the extraordinary accomplishments of the air division? How is it that the high command of the Army kept the air arm tied to the rigid framework of a land sector, hobbled to an army or even in army corps, when the functional effectiveness of the air arm required large, open spaces?

These questions may well be raised as far as France is concerned, but they would also apply to other belligerents. Though the R.A.F. was able to keep its independence it was to know many attacks and many disappointments. The Luftwaffe owed many of its early successes to the enthusiasm of a dictator enamored of all that was new and sensational;

it paid for these victories later, when it forgot the lessons of the Great War and particularly the predictable evolution of strategic bombing.

The United States had even less excuse for having failed to read the lessons of air power, since they came into the war late, when aviation had won its spurs and was winning victories as well. It is true that while Mitchell was much inspired by the British and the French, he had the greatest difficulty conveying what he had learned to America's leaders. Why was it that American generals like Pershing were not convinced by what they saw in 1917 and 1918? If they did read the lessons of air power, why did they never acknowledge those lessons?

Though this chronicle of air operations has raised a number of questions and subjects for reflection, there are many other aspects of French military aviation which would merit the researcher's attention. One case in point is recruiting or personnel procurement. Here there are a number of preconceptions, for example the tradition that aviators came from the cavalry, where the men, deprived of their horses and confined to the trenches, dreamed of combat in which the individual could distinguish himself. In point of fact the social origins of the aviators seem to be very diversified, but a serious sociological study is lacking.

What led men to seek service in the air? A number of answers have been put forward: patriotism, the desire for glory, a love of flying, and a desire for better living conditions than in the trenches. It has even been suggested that men entered the air service seeking a "safe" place in the war. This seems difficult to believe, for of 18,000 flying personnel, 5,500 were killed -- a proportion far from negligible. This question of motivation could be answered in part by using oral history techniques, for the Service Historique de l'Armée de l'Air has a large number of recorded interviews. These must be used with caution, and fortunately the conclusions they yield can be compared to the testimony of diaries and journals.

Once the airmen were recruited, how were they trained? Schools developed as they were needed, but the process did not always proceed rationally. A good part of training took place at the front, where for example observers taught themselves to pilot the planes, and we know of one pilot who got two victories at the front before finally getting his pilot's papers in 1918.

Then there is the whole question of daily life in the units, not at all the "chateau life" sometimes depicted. If the flamboyant lives of some of the aces were the exception, it seems certain that in the flying units there was more informality and less rigid discipline than in the Army. Relations between superior and subordinate were more informal, and there was a casual attitude toward dress which shocked some army traditionalists.

The subject of the aces as phenomena of the war needs study, beginning with the procedure by which they gained this title. In France the appearance of the ace made the fighter units the most prestigious and the most valorous, to the neglect of other types of air service. Was this the case in other countries? Perhaps the desire for this coveted title led to changes in tactics, with less concentration on protecting observation planes. Then in general the French aces are not well known to foreign readers today, especially to British and American ones. Much has been written of Guynemer but little of Fonck, perhaps because the latter did not meet a hero's death but survived the war by many years. If the aces were well known in their own time, there are other categories of personnel which have remained in deep obscurity; little is known, for example of aviation mechanics.

The spirit and the patriotism of the escadrille could be found in its symbols and emblems. All of these are well known, but there is an interesting field for research in studying their evolution and in analyzing and comparing them in conjunction with the practices which developed in other military organizations. The evolution of such symbols up to the present time would also make an interesting study.

The feeling of solidarity in the escadrilles found expression in their insignia, which were not brought in by the Americans, as is often stated. The insignia themselves are well known for the period of the Great War, but a study of their origins and in analysis of their symbolism could yield interesting results. Flags and unit standards could be the objects of similar studies. Insignia, flags and unit citations are the stuff of tradition, all the more cherished by the air units since they had no long history. While the customs and traditions of various units have been recorded, there is no general study of these phenomena.

The escadrille could not function without aircraft and here is an extraordinarily rich field, for materiel was in constant evolution. Victory depended not only on the quality of aircraft, but also on the quantity which could be supplied. Here there were inevitable differences between authorities at the front who committed the planes, and the authorities in the rear charged with supplying them, and of course the manufacturers. These difficulties tended to be glossed over in the aftermath of victory, but they need to be studied, using sources such as the records of the Ministry of Finance. Here too, there is a record of rapid evolution. In August, 1914, monthly production capacity was 50 aircraft and forty motors; four years later it had risen to 2912 aircraft and 4320 motors. The Service des Fabrications Aéronautiques, which was responsible for this massive increase, also strove successfully to standardize aircraft types and reduce them in number. A Technical Inspectorate assured the high quality of the products; in this respect France probably led the world in 1918. When the war ended French military aviation no doubt stood at its zenith with 3737 aircraft in service, as against 2690 for the British, 740 for the United States, and 2000 for the Germans. But if the end of the war marked the supremacy of French military aviation, it also marked the beginning of its decline.

MILITARY AVIATION FROM 1919 TO 1939

On the 11th of November 1918, French military aviation comprised 299 escadrilles equipped with the best aircraft of the period. While the performance of those aircraft still would not permit some types of use, such as long-distance strategic bombardment, further evolution and expanded use of the air arm were clearly about to occur. French aviators had at the same time acquired an immense prestige both at home and abroad, and they looked forward to continued service in the defense of their country. Yet the most immediate problem was one of demobilization.

DEMOBILIZATION OF PERSONNEL

For ground personnel the problem was a simple one to resolve; most headed home, taking with them skills and experience they could put to use in civilian life. For flying personnel the question was a more complicated one. Most of the officers who served in the air arm during the war were reservists. Some of them wanted to pursue a career in military aviation -- this was notably the case with students who had been mobilized before they had finished their studies. But many others -- bankers, architects, lawyers, and even a famous boxer (Georges Carpentier) -- wanted to return to their former professions. Their departure was a serious loss for the air service.

As it worked out, and the officers' list shows this very clearly, in the 1920s the air service had a sizable number of officers of higher rank, and a lesser number of captains, who had come from the national military schools. On the other hand nearly all of the lieutenants had come up through the ranks. The number of officers trained for staff service was very small.

This situation would be heavy with consequences all through the interwar period. First of all it made relations with the other services somewhat difficult, since for the General Staff "pedigree" often counted more than competence. Yet by the very nature of things the best aviation officers were flyers with war experience, who rarely held a rank higher than lieutenant-colonel. A second problem appeared in supplying officers for air units. Some officers of high rank had transferred from other arms and knew nothing about airplanes. Those who did have the technical competence to direct those units did not have the rank (we might note here that in France the conferring of temporary ranks in peacetime was rarely done).

There was another problem in the age distribution within the officer corps. The lieutenants who came out of the war were not very old in 1920, but there was every chance that they would grow old in the rank they then held, as would much of the corps of officers. There was a promotion regulation that to aspire to the rank of captain or major one had to be in the upper half of the seniority list. With a very large number of lieutenants already on the list, young officers fresh from school had almost no chance of advancement in the air service, and this fact had its effect on the recruitment of officers.

Then too, the officers' corps was split in another way. There was one group, the least numerous, who were the "fanatics" of flying and kept their skills at a very high level; then there was a second category, made up of men who had flown in the war -- sometimes brilliantly -- and lived to tell the tale. Now they were disinclined to run too many risks, and were encouraged in this view by chiefs who had taken to administration and felt the service might be better run without the planes.

To be sure, these difficulties were not clearly understood at the time; there was a tendency to blame the period of transition, rather than any structural fault. Up to the present very few studies have been made of the evolution of the corps of officers in the air service and the consequences of that evolution. By the use of the ANNUAIRES or yearbooks it should be possible to determine the sociological characteristics of those who left the air service by 1920 and those who stayed. It would also be interesting to see if the phenomena we have noted were peculiar to the air arm or characteristic of other services as well. What is more, relying on written or oral sources, it ought to be possible to reconstruct the attitudes, the mindset, of those aviators of the early twenties.

THE REDUCTION IN MATERIEL

Through 1919 to the end of 1920 the air service went from 90,000 officers and men to less than 40,000, and from 258 escadrilles in France to 119; the service consequently no longer needed the 11,000 planes which figured on its inventories at the end of the war. The first thought was to sell them, and French civil aviation concerns bought military craft to equip their fleets. Some fighters were bought for conversion into racing planes and attempts at records. Some of the safer ones like the Caudron G 3 were sold to the public for pleasure flying. But very soon it was obvious that the market in France would not absorb the surplus, so the government tried abroad. Such was the reputation of French military aviation that many countries asked the aid of a French military mission in setting up their own air services. Quite naturally contracts for the sale of French planes ensued in many cases, with the most widely sold type being the Breguet XIV.

Measures were also taken to stockpile flying materiel, and these stocks satisfied the replacement needs of the air service for some years. This meant, of course, a virtual end to orders placed with French builders, so the aircraft industry, which employed 183,000 people in 1918, carried only 3700 on its payrolls two years later. At the same time personnel in the technical bureaus dropped from 4000 to 40.

The consequences of this massive reduction were not seen immediately, but later it became clear that French aeronautics, deprived as it was of technical research, as well as of new planes and motors, was losing its sharp edge. We may well ask if the inability of French engine designers of the 1930s to build powerful motors cannot be traced to this inactivity of the 1920s. France had the best aircraft motors in the world in 1918, and some of the least satisfactory in 1939. In this field of materiel there are a number of studies which need to be made, both of the French aircraft industry and of its sales. It would be interesting to know who bought, and more particularly who the foreign customers were, and what economic and diplomatic ramifications -- if any -- there were in the sales abroad. What were the implications here of the government's legislation concerning war profits? Were the provisions applied fairly? Is there any truth to the allegation sometimes made that this legislation cost the industry money it would otherwise have put into research?

PEACETIME ORGANIZATION

As the air service was being demobilized, it was also being reorganized. We will not dwell on the reform of the central administration, where the essential organism remained the Direction de l'Aéronautique. The Direction lost control of civil aviation, which was fair enough, but also its purview over the industry, and this was an error. This competence lost by the War Department passed to an Undersecretary of State for Aeronautics and Air Transport, who was for a time under the Minister of Public Works, then under the Minister of Commerce. In truth the Undersecretary had the greatest difficulty coordinating the various aspects of aviation, while the military aspect escaped his control entirely. Aviation concerned several other ministries, including that for Colonies, and for Postal Service, Telephone and Telegraph. This lack of coordination would become an argument for the creation of a true air ministry.

The peacetime air forces were revamped and designed to accomplish a primary mission: supply the air cover necessary for mobilization and the opening operations of the land army. For this purpose they were grouped within an air division which would be ready at all times to carry out its mission without mobilizing reinforcements. In 1920 this division contained:

1 fighter brigade, with 2 regiments, each of 3 GROUPES of 3 escadrilles;

1 day bombing brigade of 2 regiments, one composed of two GROUPES and the other of three (each GROUPE containing 3 escadrilles);

1 night bombing brigade of two regiments, each with two GROUPES (3 escadrilles per groupe).

Another brigade was also to be established with both fighters and bombers, but of this only a fighter regiment was created, with three GROUPES of three escadrilles each. Seven observation regiments were stationed in France and North Africa, making a total of 49 escadrilles, to which should be added 8 escadrilles in the Levant and 3 colonial escadrilles. To these units should also be added supply and repair establishments and schools.

In connection with the schools should be mentioned an interesting experiment -- not wholly successful -- of flight training subventions accorded to volunteers the year preceding their scheduled military service, enabling them to acquire a civilian pilot's license. We might also note as parts of the air service seven battalions of balloonists and five anti-aircraft regiments.

The organization of the 1920s is interesting because it shows the key role the high command assigned to aviation in the beginning of a conflict. The units lived on their fields, with theoretically a field for each regiment; the regiment, essentially an administrative unit, might not be commanded by an aviator. The GROUPE directed the training activities carried on by the individual escadrilles, which remained at the base of the system and were accorded considerable autonomy in operations. Regimental strength was about a thousand officers and men, and each escadrille numbered 10 to 15 pilots, 2 or 3 of them usually being officers. A good many units were stationed outside France, with 20 escadrilles doing occupation duty in Germany and 18 in Morocco or Syria. Of the units in France, about a third were along the eastern frontier.

We know a good bit about life in the escadrilles, the sort of work which went on there, the ties of comradeship, and the hierarchical relationships which developed. We know much less about how air units got along with neighboring ground units, just how the aviators fitted into the military community or into the life of the locality in which they were stationed.

OVERSEAS OPERATIONS

This "peacetime" air service continued nonetheless to make war, for in the 1920s France was faced with revolts in both Morocco and Syria. In the latter area, where we held mandates from the League of Nations, there were upheavals in Cilicia provoked by bands supported by Kemal Ataturk and his forces, while the desert tribes of Emir Faisal were still stirred by the spectre of Lawrence of Arabia. Then in 1924 a revolt among the Druses created such problems that it took the French Army a year to crush the rebellion and bring about a period of calm which would last until 1941. Aviation played a considerable role in military operations, accompanying mobile columns and protecting their security by constant surveillance and if necessary attacks against ground objectives. The reaction of the Druse mountaineers was ineffectual, though their rifle fire could strike planes and sometimes bring them down.

In Morocco the situation had been calm all during the war; order continued to prevail there under the Resident-General, Marshal Louis-Hubert Lyautey. This stability was eventually disrupted by an uprising of Berber tribes led by Abd el-Krim which began in Spanish Morocco in 1921. The rebellion spread, and it became necessary to send sizable forces to Morocco under Marshal Pétain, who was able to best Abd el-Krim in 1926; but the war smouldered on in the Taza "pocket" until 1927, and in Tafilalet until 1932. The air service in Morocco, grouped under the 37th Regiment, took part constantly in observation and escort missions with the infantry, and served as liaison with isolated posts. In certain regions that were imperfectly known, Tafilalet, for example, Farman F 60s carried out photographic missions which yielded aerial maps, demonstrating the airplane's capacity for such work. Another type of aerial operation particularly appreciated by the ground forces were the medical evacuation flights, which saved the lives of many wounded soldiers. Despite the low level of armament of the rebellious tribes, these flights were not without risks, especially in mountainous regions, where the rebels could fire down on their targets. The 37th Regiment had significant casualties, among them two escadrille commanders, Captain Mézergues and Lieutenant Hennequin, both of whom were killed during operations.

At first glance the aerial operations in Syria and Morocco might seem to have little importance, but they impressed the aviators who took part in them and the ground forces that witnessed them. Both groups of men could see how the two forces complemented each other in the field, and how important it was for the cooperation to be smooth and well directed. But other notions were born as well from these operations. They were carried out in unusual conditions, giving only a very incomplete idea of what the work of the air force would be in a future war. There were no fighters, for there was no need for any. Bombing was carried out in difficult conditions, most often by single planes operating over mountainous terrain and using bombs of ten kilograms. Officers of the ground forces who witnessed air operations tended quite naturally to note only the value of the airplane for observation or liaison on behalf of the land forces. This tendency was all the more dangerous since so many officers of the French Army took part in the Moroccan campaign. The failure to appreciate fully the capacities of the air force that was so evident in military circles in the 1930s may in part stem from the Moroccan experience. This question would certainly bear closer examination.

MILITARY AIRLINES, "RAIDS" AND TRAINING

In France itself there were air activities which must also be mentioned. For one thing, the disorganization of the highway system following the Armistice was so great that military air lines were developed to bridge the transportation gap. A number of these lines operated from February to June, 1919, undertaking some 750 flights and carrying 11,000 kilograms of mail. The Paris-Mulhouse line, which operated a night service, deserves special mention.

Another activity, and one which occupied most air services into the 1930s, was the "raid," or long distance flight. A number of these flights were designed to tie France to portions of her empire, and such flights brought fame to aviators like Vuillemin, Dagnaux, Pelletier d'Oisy, de Goÿs, etc. Other raids were in the nature of exploits or records: Costes and Rignot, Costes and le Bris, Lamaître and Arrachart. Still other flights were designed to test new means of navigation, such as the "radio" flights of the Gérardot-Cornillon team. Another type of exercise was the aerial "tour de France" of the 1920s. Today this would scarcely be a challenging enterprise, but in 1924 it was still an exploit to be noted.

All of these flights, whose military value was not well understood at the time, served at least to perfect the aviators' skills and to bring them a certain prestige and standing with the public. The number of hours pilots flew increased steadily, but flight time was not evenly distributed. The monthly average for a pilot was 100 hours in 1924 and 120 in 1929, but certain pilots flew only the minimum necessary to receive their flying bonus, while others flew at every opportunity. Moreover flight training varied from one unit to the next, and too little attention was paid to matters of safety. It was not until 1929 that the creation of a central air security service brought clear programs in this field. During the 1920s accidents were frequent. There was a certain risk because the materiel being used was subject to breakdowns, but lack of skill or carelessness on the part of pilots also caused a number of deaths. Risk-taking and death were two themes frequently evoked in the period between the wars and oddly enough that tradition continues today. But these two themes had their positive aspects; they served to foster the idea of cooperation and mutual help among pilots, and also to bring about an improvement in the material conditions in which pilots did their work. They led to the law of 1928 which created flight pay and an air service accident fund. Later two orphanages appeared, one for girls and one for boys, to assure the education of the children of those aviators lost to the hazards of their profession. All of this still exists and the children of "those who fly" -- whether orphaned or not -- have come to know these institutions loyally supported by today's Armée de l'Air.

We might describe French military aviation of the period 1920-1928 as stable in its organization and somewhat improved in its performance over the period of the First World War. After a certain pause the French aircraft industry resumed its activity and seemed relatively prosperous. In truth the elaborate administrative procedures of the day slowed the appearance of new types of aircraft. Thus a plane called for in 1924 would be designed in 1925, built in 1927, tested in 1928, and finally enter service in 1930, though the actual time necessary to build it was less than a year. Moreover the builders were little regulated and set their prices almost at will, the government agreeing to a profit of ten percent. The builders had no incentive to build economically. The number of types of aircraft ordered was excessive; in 1927 no fewer than twenty-two new models were on order. The technical improvements being introduced were few. In 1920 the planes in service were the Nieuport 29 fighter, the Breguet XIX day bomber, and the Lioré et Olivier 20 night bomber. For observation there was the celebrated Potez 25.

ORGANIZATION AND STRUCTURE

The law of 1922 had given to aviation the status of an arm, on an equal footing with infantry, cavalry, artillery and engineers. The law of 1922 changed nothing in the organization of the air service, save that it made antiaircraft units a part of the artillery -- which helps explain why they scarcely existed in the French Army of 1940. At the head of the new arm was its INSPECTION, confided to Marshal Émile Fayolle, who was assisted by a group of brilliant young officers -- a circle which Mitchell joined whenever he came to Paris.

While the central administration remained unchanged, the structure of the force was modified in a way more satisfactory to the theorists who favored the air arm. Henceforth the air force was composed of:

1. The First Air Division (active) capable of immediate intervention and composed of a brigade of fighters and a bombing brigade. This division contained reserve units activated at the highest levels of command.
2. The Second Air Division, comprising mobilized elements formed into two brigades, whose mission was one of cooperation with the land armies.
3. Three mixed brigades not organized in divisional form.

This establishment in its entirety contained within France and occupied Germany 104 escadrilles, 30 of fighters, 32 of bombers, and 42 for observation. While this arrangement proved satisfactory, the high proportion of observation units and correspondingly limited numbers of fighters and bombers indicated that ideas on the employment of aviation had not progressed very far. While the emphasis remained on observation work, with energetic leadership the Air Division could still hope to fight and win the opening air battles, drawing upon all or part of the mixed brigades to aid in that fight.

We have now broached the key issue, the great air battle destined to win the air superiority indispensable to all other actions, be they of land or air forces. This superiority was to be obtained through the direct actions of the fighter and the indirect action of the bomber. This goal of preliminary air superiority had been demonstrated at Verdun, and became an objective in all subsequent campaigns.

Army leaders were upset by the very mention of an air battle. For them the only battle France had to fight and win was the battle on the ground. Aviation had a role in this battle, namely to observe the battlefield and furnish data for fire control of the artillery. The priority task of the fighter was close protection of the observation plane. As for the bomber, it was inaccurate and ineffectual, an expensive toy which might, in certain cases, be allowed to intervene on the battlefield.

As far as the aviators were concerned, there would undoubtedly be an air battle at the very beginning of the war. What is more, it seemed a necessary consequence of the theory of the inviolability of fortified fronts, another dogma professed by the French General Staff and the Ecole de Guerre. Since the opposing land armies were obliged to confront each other without piercing the opposing line, only air actions above the lines would be capable of inflicting sufficient harm to the enemy to make him give up the conflict. This thesis, essentially that of Douhet, proceeded very logically from the affirmations of the General Staff. In spite of what has been written, Douhet's ideas were not completely shared by French airmen, who did not accept his views on the results of long-distance bombardment missions.

Yet land army spokesmen took up these very notions to portray a sort of apocalypse which waves of enemy bombers would bring to France, bringing to her cities an aerochemical war which would destroy all life within them. This argument registered both in French political circles and in public opinion. Grass-roots France was still haunted by war, and the simple evocation of these terror raids was enough to bring about a condemnation of aerial bombardment altogether. Then too, the bomber was an offensive weapon, and France sought only to defend herself. This sort of argument seems facile today, but it carried considerable weight; ultimately it even affected the designation of air units, so that the fighters were rebaptised "light defense aviation" and the bombers "heavy defense aviation."

What lay at the base of these unfortunate quarrels, which became more intense when French military aviation became an independent arm? It might be seen as a simple clash of opposing viewpoints, each side holding firmly to the rightness of its own position. But what further complicates the question of motivation is the refusal to read the lessons of the previous war. For aircraft, as for tanks, these lessons were clear -- and they contradicted the official teachings of the Ecole de Guerre, the French Army's doctrinal authority. It is noteworthy that in the same period British and American armies were imbued with similar notions, and subsequently their thinking would be contradicted by events. It is interesting to speculate whether these common positions were a logical synthesis of the various studies and analyses which were made after the Great War, or if this common thinking in the Army circles of several countries was a sort of reflex of established military institutions faced with new and dynamic competitors.

And behind the defense of certain conceptions was revealed the defense of interests: the established order of things, the past and prestige associated with it, and status and position, which the old institution was loath to share with the new. There is a fascinating field for study at the comparative level, where an examination of several countries might provide a composite portrait of military conservatism as it faced the new challenge from the air.

The doctrinal debate, which went on until 1939 and was resumed in 1946, led many airmen to conclude they would never convince higher Army authorities of the rightness of their views -- so the only means of creating the sort of air force they sought was to change the position of subordination to the Army high command, a subordination which brought all their efforts to naught. In a word, it was time to elevate military aviation to the status of an independent service with its own superior authority, an air minister. We shall see later that many arguments were advanced in support of independence, but for the moment let us confine ourselves to the airmen themselves and their motivation. Was their desire for independence unanimous and sincere? It was certainly not unanimous, and here again a study of the corps of officers would be useful in giving a precise answer. Almost all of the officers of higher rank then serving had come from the army, and many of them had spent a large part of their careers in the land service. Some of the higher officers did not fly, others did so rarely. For many of them the new service did not offer the same career advantages as the army itself. But obedience and respect for hierarchy kept officers from expressing opinions contrary to the official view, so that research into those opinions and the reasons for them is not easy: though "independence" was a burning issue, airmen did not often express their opinion on it and when they did, it was often under the veil of anonymity.

We might also raise the question as to whether the airmen played a major role in the creation of an "all the way" air ministry; the answer would not be straightforward. Their role was probably not direct but certainly it was indirect. The arguments they put forward in their own circle may also have been expressed in meetings with others in the aeronautical world; with aircraft manufacturers, for example, they probably had occasion to discuss the future of military aviation.

That future was a matter of some concern. In May, 1926, the position of Secretary of State was abolished and his functions distributed to several ministries. Coordination became very difficult, producing catastrophic repercussions in the aeronautical industry, which passed through a severe crisis. And this crisis produced a technical lag, evidenced by the loss of world records between 1925 and 1927. The legislator who put forward the aviation budget for 1928 began his presentation with these words: "French aviation is in a very perilous state." Then too, the defenders of aviation found it disconcerting that almost by whim it should be passed from one ministry to the next: "Yesterday under Public Works, today under Commerce . . . tomorrow . . . a satellite of Agriculture." It was likewise shocking that an activity of such importance should be classified simply as an administration, like the administration of the government's stud farms.

A concordance of interests and concerns pointed to the immediate need for an air ministry. Unfortunately there was a major obstacle in the person of Maurice Bokanowski, Minister of Commerce, who wanted to retain control of air transport. A fortuitous event removed this difficulty: M. Bokanowski was killed in an airplane crash on September 2, 1928, and on the 14th the Premier, Raymond Poincaré, created a Ministry of the Air which he confided to M. Laurent-Eynac.

With the creation of the ministry, there was an enormous amount of work to be done. Monsieur Laurent-Eynac had to try to resolve the crisis in the aeronautical industry, make an effort on behalf of commercial aviation, and then do the legislative groundwork for the creation of an independent air force. To accomplish these things he had to overcome considerable resistance, and notably the misgivings of the Army and Navy general staffs, now shorn of their cooperation aviation, in which in truth they had never shown a great deal of interest. From 1928 to 1933 the War Department opposed an independent air force, without however greatly hindering legislative activity in this regard. The Navy, on the other hand, blocked all action until a 1932 agreement between M. Paul Painlevē and M. Georges Leygues secured to "la Royale" (the French Navy) control of maritime aviation. It is striking that in a continental nation like France, it should be the Navy which most effectively resisted the creation of an independent air force. The Royal Navy played this same role on Britain, though with less success against the R.A.F., but the U.S. Navy successfully developed its own air arm in order to combat certain ideas developed by Billy Mitchell.

It would be enlightening to trace out the forms taken by the opposition, which was a breach of discipline, since it challenged a governmental decision. Army and Navy directed their efforts quite differently. The Army had partisans in the new ministry; General Emile Hergault, for example, Chief of Staff of the Air Force from 1931 to 1933, was noted for being attentive to the Army and anxious not to trouble his friends there. Quite soon the Army high command gave up the clamor for returning its air units and set its efforts toward getting an air operations manual which conformed to its views. In any case the Army could hardly ask for its units back, since their transfer would gut the new ministry, and this the government would not permit. For naval aviation, which was numerically less important and indissolubly tied to the Navy in the case of its units on shipboard, the problem was an entirely different one. Naval aviation could be returned to the Navy with scarcely any damage to the principle of a "united" air force. For the historian this transfer offers an interesting lesson. One of the key arguments of those who opposed an independent air force was the assertion that an air force which remained part of one of the older services would have a more promising and harmonious future. The poor state of the Aéronavale -- the French Navy's air arm -- in the critical year 1939, and the Navy's cancellation of two modern carriers which were to replace the old BEARN, point to the fallacy of this argument.

Other sources of opposition are less clearly tied to the Army and Navy "lobbies." Some of the ideas advanced were surprising at the very least. In the prestigious REVUE DES DEUX MONDES a writer identified as a highly qualified personality claimed that an independent air force would be deprived of "the traditions of discipline, of moderation, and of political reserve of the two old forces of land and sea." By giving independence to a professional force, a force with cadres "in which the taste for thrills and risks is general, an army composed of individualists not always easy to subject to strict discipline," France would be making her air force "an army of coups d'état par excellence." This text, which one deputy claimed would provoke general hilarity in the escadrilles, makes specious arguments based on the opinion of the older services, who saw airmen as daredevil adventurers and individualists incapable of discipline.

The new Air Minister faced other problems more serious than these polemics. The first task was to resolve the crisis of the aeronautical industry. To this end he created the Direction Générale Technique, which was confided to M. Albert Caquot, an engineer of exceptional ability who had worked with aeronautics during the war. The minister also recognized that developmental work had been sacrificed in the interest of serial production, that manufacture was scattered among too many firms, and that the stocks of planes and motors set aside for mobilization rapidly became outmoded. The Minister launched a new policy designed to use more funds for development, to spur concentration in the industry and to substitute industrial mobilization for stockpiling.

M. Caquot implemented this policy. Rather than build outmoded planes, he called for the construction of the prototypes of aircraft of superior performance. This progressive policy, known as the policy of prototypes, created a storm of protest, particularly on the part of manufacturers unhappy at the appearance in their industry of young engineers anxious to attempt something new. Others claimed that some of the new aircraft developed had poor flying qualities or were even dangerous. In actual fact the policy brought some excellent results, bringing out some very fine aircraft and encouraging dynamic manufacturers such as Emile Dewoitine and Marcel Bloch. The abandonment of this policy in 1932-33 in favor of "transition" aircraft in large numbers, ensured that the Air Force would have outmoded planes, and from 1936 deprived it of new models to build.

Today we might say by way of criticism that this policy entrusted initiative to private enterprise, rather than challenging it with ambitious programs. What is more, the other plans of the ministry were not realized. It did not receive the funds necessary to prepare industrial mobilization, while manufacturers themselves showed little interest either in concentration or decentralization. This state of affairs helps us understand the purpose of the nationalizations of 1936, which were the only way to impose upon private builders the changes necessary for the common good.

From 1930 to 1932 there were a number of organizational changes, but only those of 1932 made any fundamental difference to the pre-existing structures by setting up experimentally an organizational plan already laid down in a scheme proposed in 1929 but never enacted. This change produced the definitive unit structure of the Armée de l'Air by replacing regiments with ESCADRES and fixing the basic air force unit as a GROUPE composed of two escadrilles. (The ESCADRE corresponded roughly with the American Group and the British Wing; the French GROUPE was equated with the British Squadron, and the escadrille with the flight.)

Another change, one which stirred up a tempest of controversy, was the adoption of a new uniform. It was navy blue with an open collared coat, topped oft with a cap and for a symbolic arm, a dagger. Thus it broke completely with the tradition of the Army, which considered the airmen had committed a sort of sacrilege in giving up the képi. At the same time the French Navy was not at all happy that the airmen had adopted their traditional dark blue, leading to cases of mistaken identity which the Navy felt would be embarassing. Yet the new uniform, ill suited as its color was to life on an air base, became the first symbol of the unity of the new force and was appreciated for that reason. Today it is difficult to understand all the importance attached to this change. One can only wonder if such changes led to similar repercussions in other countries, and what might have caused them.

On January 31, 1933, Pierre Cot became Air Minister in the Daladier government. Beginning on February 7 he made General Victor Denain Chief of Staff of the Armée de l'Air. When Cot left the Ministry General Denain took over that post and held it until January, 1936. In July, 1936, with the accession of the Popular Front, Pierre Cot returned to the Ministry and held it until January, 1938. We can say that except for one brief period from January to June, 1936, when Marcel Déat held the portfolio, the leadership of the Armée de l'Air possessed a considerable stability.

After 1933 and the arrival at a compromise with the Navy, there was no further obstacle to voting the organizational scheme arrived at in 1929. The first of the laws now voted gave the Armée de l'Air a structure conformable to its operational doctrine as defined by decree of March, 1933. This decree assigned to it the task of cooperation with the ground forces and also that of conducting the contest in the air and air defense. So long as these principles were only that, they were satisfactory; but when it was necessary to create a force charged with carrying them out, things did not go so smoothly. The Armée de l'Air wished to enhance its capacities in bombardment and also in fighter activities, even at the risk of reducing its forces earmarked for cooperation with the Army. This the Army would not accept. General Maxime Weygand, Inspector-General of the Army, stated that "the Air Minister's notions of the air war, the organization of the air forces, and the possible reductions to be made in observation units create the most serious concern." Somewhat later Weygand declared that "if the Air Department follows its announced plan for aerial rearmament . . . , the War Department will be obliged to organize with its own means the Army cooperation aviation necessary for its large units."

The last phrase is interesting. For the period up to 1938 the Armée de l'Air had great difficulties getting appropriations, while the Army declined to give up any of its funds to ensure the sort of air force it said it needed. Yet the Army would have been able to find the appropriation, if it had come to that, to create its own support aviation. Something of the same phenomenon occurred a few years after World War II, when the Armée de l'Air, critically short of funds, had to accept the creation of a light aviation unit within the Army, financed by Army funds. Here we may get some idea of the budgetary process and how wasteful it could be. It would be worth the historian's while to probe this matter further.

The law on organization of the Armée de l'Air enacted July 2, 1934 stipulated that in time of war the nation's air units would be divided into two parts, "reserved" forces and those destined for cooperation with the Army. But this arrangement was not valid in peacetime, where organization was territorial, based on air regions -- an old idea laid down in 1873, affirmed in 1882, and further sanctified by the law of 1927. As a consequence, at the moment when a war broke out, the time when it was important for the Armée de l'Air to "cover" mobilization, it would itself be in a state of transformation. We can only imagine the confusion which would have reigned, had the Germans launched an attack against us on September 3, 1939. Luckily this was not the case, and between that day and May 10, 1940, the various commands had plenty of time to establish themselves.

At the same time this reorganization was under way, the Air Minister also decided upon a restructuring of the central administration which would strengthen the position of the Air Staff. Two other important institutions were also created at this time, the Ecole de l'Air and the Ecole Supérieure de Guerre Aérienne. The Ecole de l'Air was to become the equivalent of the Army and Navy academies. Created at Versailles in 1935, and installed at Salon de Provence in 1937, it began to form generations of flying officers, though it was not until 1969 that a graduate of that institution became Chief of Staff of the Armée de l'Air. The Ecole Supérieure de Guerre Aérienne (Air War College), had as its function to mold future air leaders, giving them a program of study more suitable than that of the Army's Ecole Supérieure de Guerre. By these measures General Denain thus fleshed out the structures of the Armée de l'Air before leaving the Air Ministry.

THE BCR

General Denain left another legacy, a long-term development plan which contained the highly controversial BCR program, which was to supply the Armée de l'Air with a bombardment-combat-reconnaissance aircraft. This multiple purpose plane had antecedents in the Great War, but in the 1930s it had a specific goal to give the Armée de l'Air the bombers it felt it needed, and the army cooperation planes it required, by combining these functions in a single design. In December, 1935 a General of the Armée de l'Air defined the function of the BCR as follows: "its role is not really to seek air combat, but it should permit reconnaissance units, the most numerous in the service, to reinforce the heavy aviation [bombers]". This program spawned the Bloch 200, the Amiot 143, the Bloch 210, and the Potez 540. Alas, none of these planes was capable of daylight operations in enemy airspace in 1939, though Amiot 143s would be committed over Sedan in May, 1940. This handicap marked all the aircraft of Plan I of 1933.

This Plan I, also called the 1016 airplanes plan, was designed to provide 1016 aircraft: 350 bombers, 350 fighters, and 316 reconnaisance planes. Issued in 1933, it specified production of aircraft already developed, the fighter designs of 1930 and the bombers of 1932. The plan, to be completed in 1936, had no built-in modernization feature. The Armée de l'Air thus found itself equipped with planes outmoded almost as soon as they were built, while the industry was incapable in 1936 of furnishing more modern replacements. Some believe that this scheme was devised by General Denain in the belief that France would have a war with Germany in 1935 or 1936, the period when Plan I would peak. But in the crisis of March, 1936 the French government declined to call Hitler's bluff, and perhaps on that day France lost the war which lay ahead.

By 1937 comparisons with the Luftwaffe's materiel were alarming, and the government of the Popular Front moved to redress the balance. Pierre Cot returned to the Ministry. While his return was not universally welcomed, and while he was pilloried in right-wing publications, we should note that career officers in the Armée de l'Air did not vote or much concern themselves with politics. The Service Historique has interviewed officers concerning their attitudes in that era. The results show considerable variation, with loyalty to the legal government a strong feature of military society. These attitudes merit more thorough investigation by historians.

In the summer of 1936 the government had to turn its attention to the problem of increasing aircraft production, and a program of nationalizations of aircraft firms was decided on in August of that year. This action has been much criticized, but in 1936 it was the only way to achieve the concentration and decentralization which had been sought in vain. Nor was it a radical type of nationalization: the government simply became majority stockholder in the companies involved. Management and research and development would often continue as before.

The nationalizations have been blamed for the disorganization in our aircraft industry, for the many strikes which plagued it, and consequently, for the defeat of 1940. It is true that there were difficulties in carrying out the nationalizations, and it is true that they provoked much disorder. But how else could one hope to correct in a short time the errors of many years? Then too, much of the criticism of nationalization is really a criticism of the Blum government, with its social legislation, its forty-hour work week and its paid vacations, which many see as hampering the war preparations France needed to make.

Objective studies on this subject, as well as on the reported sabotage in armaments factories, have yet to be made. Even today, for most French scholars these topics still have political or ideological overtones, so that the results of their studies would be challenged by one side or another. Perhaps a foreign scholar's work might be more readily received here.

REARMAMENT PLANS

The reorganization of the industry was followed by new rearmament schemes. Plan II, which appeared in September, 1936, was an overhaul of Plan I, designed to produce 2795 planes by the end of 1939, 1518 of them first line machines. This plan, which unfortunately led to orders for outmoded aircraft, gives us some idea of the priorities envisaged by the leaders of the Armée de l'Air. Bombers were to be 45% of the planes produced, fighters 25% and observation/reconnaisance craft 30%. This plan also introduced units of paratroopers, a type of soldier which the land army found hard to accept. We should also mention the development of Plan III, destined to renovate antiaircraft defenses. A Plan IV was proposed for light defense aviation (fighters), but it was not put into effect, instead came Plan V which we will treat further on.

PERSONNEL PROBLEMS

Though the legislation of this period chiefly concerned materiel and the aircraft industry, there were problems in procuring qualified personnel in sufficient numbers, these being largely attributed to budgetary restrictions. There is a subject for serious study in the attitudes of the rue de Rivoli (that is to say the financial organisms of government located there) towards defense spending. The shortage of personnel which resulted from financial stringency was paralleled by a decline in quality. Flying personnel were kept flying too long, and some had difficulty adapting to more modern aircraft. There is something of a revolution in flying between 1930 and about 1937, with better performance, flaps, variable pitch propellers, retractable landing gear and techniques of bad weather flying, all of which were a redoubtable challenge to older pilots. A remarkable training program for blind flying, introduced in 1936, was chiefly successful with young pilots. Lack of training plagued bombing units above all, while fighter pilots steadily improved their skills under programs of instruction introduced by General Massenet de Marancour and his successor, General d'Harcourt. This training enabled French fighter pilots to meet the challenge of the Luftwaffe despite their inferior numbers.

There were notable shortages in two types of personnel: mechanics and pilots. In the case of mechanics recruiting was hampered chiefly by the low pay. The numbers of pilots could be increased either by enlarging the classes at the Ecole de l'Air or by preparing reservists who could then be moved to active duty, but both of these solutions took time. There was a scheme for increasing the number of reservists by means of "Popular Aviation," a subject still largely unstudied. In order to enable young men to perform their military service as pilots, in 1922 the government created pilots' grants, covering the cost of flight training in civilian centers. The number of grants was small, for it was an expensive form of training, and even more had to be allocated to keep up the reservists' flying skills. Pierre Cot enlarged upon this conception with his "Popular Aviation," which generated great enthusiasm for flying among the French public, though it came too late to really build up the pool of military pilots. Cot introduced further reform by lowering the permissible age for flying personnel and by bringing officers up through the ranks in sizable numbers. These last two measures were heavily criticized as being politically inspired, as indeed they may have been. Cot and other political figures of the left tended to regard officers as reactionary, and non-commissioned officers as more sympathetic to ideas from the left.

These political attitudes need to be studied. How, for example, did the noncommissioned officers of the Armée de l'Air view the Popular Front? The opinions of officers are somewhat better known. They were, for example, in the great mass anti-Communist. Is this a class reaction? This has been argued, but needs to be investigated. Certainly the corps of officers was not at that time a "caste." The service academies were avenues by which the sons of humble families could go up, though here again a sociological study needs to be undertaken.

ORGANIZATION

The men around Pierre Cot saw clearly the shortcomings of the organization of 1934. Cot moved to separate operational and territorial commands, with the former now concentrated into two large units, one of them "heavy" aviation, i.e. bombers, and the other "light" aviation. In time of war, operations would be directly by the chief of an air army, composed of heavy and light units. This arrangement broke new ground and put our wartime organization on a level with that of other major powers, though it was not without its faults. These faults led Cot's successor, Guy la Chambre, to revert to the organization of 1934.

At this time General Joseph Vuillemin was installed as Chief of Staff of the Armée de l'Air, a choice that has often been criticized. Though as a hero of the War of 1914-1918 Vuillemin was well received in the air units, he had a severe handicap in that he had no staff experience. In this situation he relied heavily on his assistants, General Mendigal and General Bergeret. These officers had both taught at the Army's War College, and it has been said that their close ties with the land service led them to accept its wishes all too readily. This hypothesis has yet to be tested by historical research.

In 1938 there was great pressure to accelerate plane production, while at the same time to drop outmoded designs and retool for more modern types. All this took time, and production went well only for the Morane 406, the Potez 63, and the LéO 45. Even with these models, appreciable numbers were not expected to come from the assembly lines until the end of 1939. To help fill the void, other planes might be bought abroad. The idea was not new, for in 1937 Cot had sent a mission to the United States to look into the possibilities there. Guy la Chambre pushed on with these plans despite opposition from two quarters. There was first of all the hostility of the French aircraft producers, who wanted to supply all the government's needs themselves. Then there was the American regulation which forbade the export of military aircraft until two years after they had come into use in the American service.

Negotiations with the Americans finally led to a contract for 100 Curtiss P-36s, signed in May, 1938; a little later agreements were reached for the sale of Glenn Martin and Douglas DB-7 bombers, BT-9 trainers, and Chance-Vought dive bombers. New orders for P-36s were signed, and the first of these craft arrived in France in December, 1938. By the time of the Armistice in June, 1940, the Armée de l'Air had received 489 of the 980 American planes ordered (40 Chance-Vought, 283 Martin, 120 Douglas, 316 Curtis, and 341 of the BT-9s). While these purchases have been studied, more work needs to be done, notably in determining their affect on the American aircraft industry.

The year 1938 also saw the formulation of Plan V, approved on March 15 of that year, to be completed by March 1, 1941. In contrast to Plan II, this new scheme gave absolute priority to fighters -- this despite the official view of the Armée de l'Air that air strength should be above all offensive strength, with the bomber as its critical element. Though the Armée de l'Air never abandoned this principle, it was obliged to accept a delay in the creation of a modern bombing force, chiefly due to technical problems in the production of the LéO 45. For the time being, as the danger of war increased, the chief preoccupation became warding off the blows the German air offensive was bound to deliver.

Plan V foresaw the production of 4739 planes, 2617 of them first-line craft. Revised several times, the last figures before war came provided for 8094 planes, 2938 of them on the line. M. Caquot, placed at the head of production in September, 1938, sought to speed up production of having the various plants produce subassemblies, rather than entire planes. Final assembly would be confined to a single plant. Production stood at 40 planes a month in September 1938, and at 300 per month in August, 1939.

The increase in production was all the more necessary in that the Munich crisis of 1938 found France without a single modern plane in its air units -- General Vuillemin was to admit then that the Armée de l'Air would be powerless against the Luftwaffe. Some say Vuillemin was overly apprehensive, since the Germans had bluffed him into believing the Luftwaffe was far stronger than it was in reality. But German air power was all too real, and estimates of its strength by French air intelligence were quite accurate. In any case it was not weakness in the air which produced Munich but the reluctance of France to go to war, a reluctance her leaders understood full well. Chamberlain's government wanted an accord with "Mr. Hitler" and the French government went along.

In 1939 the situation was different. Our inferiority was evident, but it was not total. At the outbreak of war France had 20 fighter GROUPES within France itself, 17 of them equipped with Morane 406s, Curtiss P-36s or Potez 631 night fighters. The bombing force embraced 21 GROUPES, all of them with outmoded planes which could not do day bombing without heavy losses. There were 12 reconnaissance GROUPES, 4 of them equipped with the Potez 637 which had a higher ceiling than German fighters. In one domain the Armée de l'Air outclassed the Luftwaffe, and that was in observation planes. There were no fewer than 35 GROUPES, equipped with the Mureaux 115, LéO C30 autogyro, Potez 25 and 540, and the Breguet 27. Very quickly these planes proved incapable of "crossing the line" and returning and they had to be withdrawn from use.

We have dwelt on the importance which the Army placed on observation planes and the little use it had for bombers. The Army obtained control of antiaircraft weapons in 1922, and despite the existence of excellent prototypes, it was slow to put these weapons into production. Army units had virtually no antiaircraft weapons, nor did they have any training in resisting air attacks. These oversights would cost France dearly in 1940.

As for the Armée de l'Air, though it went to war in a manifest state of inferiority, it was at least in the process of renewing and rearming itself. The factories were humming and the flight schools filled. All the airmen needed was time, and that the military pundits all promised. The Army and the Maginot Line would stop the German armies, and by the spring of 1941 the Armée de l'Air would be equal to the Luftwaffe. Alas, within a few months the Germans would shatter the front, and with it a generation of illusions. The Armée de l'Air would have to fight, not for the prospect of victory, but for its own honor.

THE SECOND WORLD WAR

On the outbreak of war on September 3, 1939, the units of the Armée de l'Air had already been alerted. They moved to operational airfields on August 22. Since the Luftwaffe was for the time being heavily committed in Poland, it could not immediately launch the air offensive that had for so long been feared and expected. Though for airmen the "Phony War" would produce violent clashes -- unlike the situation on the ground -- the French air units would nevertheless have time to adopt their organization to the wartime conditions, to add to their effectives, and to replace some of their materiel.

THE COMMAND STRUCTURE

Under the orders of General Vuillemin, commander in chief of the Armée de l'Air, were three air armies. The First Air Army under Command of General Mouchard covered the theater of operations in the northeast, the Third, commanded by General Houdemon faced the Alps, and the Fifth, under General Bouscat, defended North Africa. On this somewhat cumbersome organization were superimposed the ZOA or Air Operations Zones corresponding to the land armies. Thus the Northern Zone (ZOAN) was set up in October, 1939 under General d'Astier de la Vigerie, and a little later an Eastern Zone (ZOAE) under General Pennès. These zones were under the commander of the First Air Army, while at the same time subordinated to the commander of the corresponding group of land armies. At the end of October still a third zone was created in the south (ZOAS), subordinated to the third Air Army.

These structures were far too burdensome and the Air Army level proved to be an obstacle to air command and a source of conflict with ground command. These armies were abolished at the end of February, 1940 and the forces placed in an Air Forces Cooperation Command (General Tetu); now there were four zones of air operations North: (General d'Astier), East (General Bouscat), South (General Odic), and the Alps front (General Laurens), plus an air force command in North Africa.

Other measures were taken to improve the organization of bombardment and attack aviation, but the real problem was not there. The Armée de l'Air, which took such pains to possess structures permitting it to conduct air operations with massed forces, could not carry out such operations because the air commander had practically no force reserved for his use, the fighter and reconnaissances GROUPES being, for the most part, parcelled out to the different land armies.

AIRCRAFT DELIVERIES - THE AERONAUTICAL INDUSTRY

In September, 1939 the problem in the field of materiel was actually a double problem: how to increase airplane production and how to produce modern aircraft. A key feature of this problem was the lack of production personnel. At the outbreak of the conflict very large numbers of qualified workers were taken in military mobilization, because there was no industrial mobilization which would have kept them at their jobs. If their departure disorganized the factories, the arrival of their replacements had a similar effect. The aircraft industry's newly employed workers increased from 170,000 to 250,000, among them many women incompletely trained; this was one of the chief causes of the deficiencies in materiel which were noticed in the fighting units. Such deficiencies were all the more possible because of the absence of qualified controllers, who had also been put into uniform.

Can these deficiencies also be explained by sabotage, notably on the part of Communist workers following party instructions? The question is worth the asking, but an answer is not easy. There is no doubt that there were cases of willful sabotage, and some of these were proven and punishment meted out accordingly, but it is far more difficult to tell if cases of faulty workmanship detected in the units were due to ill will or simple incompetence.

Other considerations further limited aircraft production. If in general airframe makers worked rapidly and well, motor producers did not keep up with them, and those responsible for other equipment performed poorly. There were shortages of instruments, radios, landing gear and variable-pitch propellers. Numerous aircraft apparently completed in May, 1940 could not be used because they lacked components that had not been furnished in time. This is why orders placed abroad and notably in the United States, included not only aircraft but also motors, propellers and flight instruments.

At the time France entered World War II goals had been set for the aeronautical industry which it had no chance of achieving. Production did increase from month to month, until in May, 1940 it reached 450 aircraft; they were above all modern airplanes, destined to replace outmoded craft which were dangerous for those who flew them in front line units. The increase in production and the delivery of American aircraft should have led to a renovation of front line units and also an increase in their numbers. Unfortunately -- and this is often forgotten -- the difficulties encountered in training flight personnel led to an insufficiency in trained air crew, limiting the number of units.

The greatest effort was made in fighter aviation, but the delays in bringing out the Bloch 252 and the Arsenal VG33 limited the possibilities here. Even so, the 23 GROUPES (19 with modern aircraft) which existed in September 1939 had been raised by the following May to 28 GROUPES (fifteen with Morane 406, eight with Bloch 151 or 152, four with Curtiss H-75, and one with Dewoitine 520). As for bombardment aviation, the problem was only in part a delay in deliveries from factories, especially deliveries of the Amiot 350, whose flaws the builder found difficult to correct. Then there was the problem of training crews for the LéO 45, which required particularly deft handling and represented a big change from the LéO 20 and the Bloch 200. On May 10, 1940, of thirty-three bombardment GROUPES, seven were equipped with modern craft and twenty were in process of converting to LéO 45, Amiot 354 and Breguet 693. The reconnaissance GROUPES, which had been able to fullfil their mission thanks to the excellent Potez 637, found it difficult to penetrate deeply into enemy territory because the Potez 63-11 was an easy prey to new German fighters. Deliveries of the Bloch 174 came too late to change the situation very much.

Observation aviation was destined to demonstrate the absurdity of using such craft as the Mureaux 115 and autogiros without having first obtained air superiority. Interestingly enough a recent book (H. Dutailly, LES PROBLEMES DE L'ARMEE DE TERRE FRANCAISE, 1935-1939) noted the desire of the Artillery to have autogiros for use as flying observation posts. The Artillery criticised the Air Staff for having said "the risk of destruction of artillery aviation of this type appears to be so great that it would not be able to carry out its mission." Be that as it may, the observation groups carried out their mission with the Potez 63-11. This was the case for the thirty-five GROUPES engaged on the northeastern front on May 10, 1940.

THE AIR WAR DURING THE DROLE DE GUERRE

While it is true that for the land army there was no fighting, except for small skirmishes, until May 10, 1940, for the Armée de l'Air the fighting began in September, 1939. The air action was chiefly reconnaissance and fighter activity, the bombers being used mainly to scatter leaflets over Germany in night flights. These actions were sufficient to demonstrate the failings of certain aircraft; the Bloch 131 and the light observation planes were withdrawn from operations. During this period sixty-three aircraft were destroyed and eighteen damaged; the Luftwaffe lost eighty planes (plus thirty-four probables). To be sure, the air battles of May and June would produce far heavier losses, but those suffered during the "Phony War" were not negligeable.

THE BATTLE OF FRANCE

The German attack began on May 10 with massive fighter and bomber attacks on French airfields. True to the airman's unchanging doctrine, the Luftwaffe tried to establish air superiority in the skies of France at the very outset. This superiority was already achieved in the number of aircraft. Germany had on the line 1210 fighters, 1680 bombers and 640 reconnaissance craft. France had 637 fighters, 242 bombers and 409 reconnaissance airplanes. To this force we must add the British air units in France, about 450 aircraft, with 140 fighters and 240 bombers.

Since the war any number of arguments have been advanced to show that in reality the total number of French aircraft was comparable to those of the Germans. These arguments are essentially put forward by non-specialists in the matter of aeronautics, who simply add up all the aircraft, of whatever type, including relics of the 1914-1918 war, such as the Breguet 14, and aircraft of the thirties, Potez 25, Nieuport 29, etc. It should be evident that this procedure is absurd, and that one should take into consideration only modern aircraft, ready for service. As for the struggle for air superiority over France, the key figures are those relating to fighters, and here again the figures are somewhat deceptive. To oppose 1210 German fighters, the Allies had 777 aircraft, 637 French and 140 British. Apparently the ratio was not too unfavorable to the Allies, since for each Allied fighter the Germans had 1.6. This balance of forces was further influenced by two factors; one of them was the advantage of the initiative held by the Germans, and the other was the involvement of the French fighter force in Army support roles.

The fact that Germany held the initiative had three important effects. Knowing the day of attack beforehand, the Luftwaffe was able to manage its force so as to have a maximum of planes available for May 10; the French obviously would not. As a result the proportion of Luftwaffe aircraft not available was about twenty percent while that of the Armée de l'Air was around thirty percent, the normal or usual rate for air forces of the era. Then, having chosen the points of attack, the Luftwaffe could concentrate its forces on them. The French fighter force was obliged, in its defensive stance, to spread its units over the whole of the war zone; what is more, in May 1940 the French fighter command did not have the means of detection and transmission to concentrate fighters on enemy raids. Then, finally, Luftwaffe raids on French airfields, while not as successful as the Germans had hoped, nevertheless put a number of French planes out of action.

To these handicaps faced by the Armée de l'Air we must add others which derived from the structure of the air arm. Numerous fighter GROUPES were placed at the disposal of the land armies, whose leaders had a very limited idea of how to use the aerial resources under their control. They kept on the ground airplanes which could otherwise have been thrown into an air battle which could not be confined within the narrow limits of each army's operational zone. The retention of aircraft on their landing fields explains in part the small number of French pilots who flew missions on May 10 -- fewer than 200 pilots flew 360 sorties. A detailed study needs to be made to determine which units were not engaged and why. In any event, all of these limitations on aircraft use had effects on the balance of forces which were disastrous for the Allies; in point of fact -- and this fact should be remembered -- Allied fighters had to face their German opponents in a proportion of about one against six.

Bomber units, like fighters, were dispersed, and their concentration difficult in both time and space. Even so, given the way the land campaign developed, they had to be thrown into the battle. From May 11 onwards, French bombers were sent in "penny packets" against enemy columns. The LéO 45 and Breguet 693 could then discover all the redoubtable effectiveness of the German antiaircraft weapons. These weapons, which we had neglected, were amply supplied to German ground units, and in the hands of well-trained soldiers. Neither the Flak nor the German fighters deterred the French command from sending out on a day mission on May 14 Amiot bombers whose slowness made them unsuited for daytime operations. Such operations were also bad for the fighter arm, since they made heavy demands for fighter escort, diverting that arm from more productive activities. Once the fighters were freed from this obligation they engaged the German fighters on better terms, inflicting heavy losses. It is not necessary here to enter into the details of these battles. A number of studies have been made drawing on very detailed sources kept by the S.H.A.A.

Many studies remain to be made on this period. As for the industrial aspect, we need to know how aircraft production proceeded and how replacement aircraft were channelled into the fighting units. We have already seen that many planes were produced which were judged not ready for service. Moreover there were numerous malfunctions and breakdowns, and it is fair to ask if this involved incompetence or sabotage. As for the flight crews they had no doubt about sabotage. We may also ask what effect this belief -- well-founded or not -- had on these aircrews.

The shortages of material were frequently cited in the Riom trials among the charges against aviation leaders, but were they as significant as was charged? If so, just where should the responsibility be placed, among industrialists, political figures or military leaders?

Material insufficiencies were much talked about, but human shortcomings almost never. Was this because it was difficult to fix such shortcomings, other than on the military functionary technically responsible? Many cases can be cited which had considerable consequences; that of the Ecole de l'Air is typical. In 1939 the Ecole de l'Air was installed in Salon de Provence, where there was good flying weather, except when the mistral was blowing. But since the location seemed threatened by the near proximity of the Italian frontier, the institution was moved to Bordeaux, where flying conditions were poor in winter. Why this decision and why not a move to North Africa, where the weather conditions are particularly favorable for flying? Why was the training in this school so slow? Lack of teachers, of aircraft, of doctrine? Such questions still do not have satisfactory answers.

Research efforts need to be directed toward the general costs of this campaign. On the French side the cost is known. The losses were heavy, since a quarter of the flying officers with the armies were killed. The number of aircraft destroyed was also considerable: more than half of the fighters and bombers were lost. To these figures must be added the number of planes temporarily put out of action by the intensive effort fighting units made. The flow of replacement aircraft, while far from negligible, could not make up for the losses. Then too, the losses in personnel were not fully replaced. A few pilots fresh from flight training and some flight instructors went to the front to fill some vacancies, but the relief they afforded was only temporary. The great majority of flying personnel were physically exhausted, and the fact that they fought on when all hope was gone gives us some idea of their mettle.

French aircraft losses include those shot down by the enemy, those destroyed on the ground by enemy bombardment, and those lost through accidents. For the Luftwaffe the losses are difficult to establish. For the time being we have only the French estimates, which include only aircraft brought down by French fighters or anti-aircraft fire. These losses are estimated at 853, about a quarter of the German front line aircraft on May 10. Losses due to accidents should be slightly lower; these were none caused by using bomb damaged airfields, since the Allied air forces did not carry out such bombardments. But in time of war, in a period of intense effort, the number of accidents is never negligible; these result in part from damage inflicted on the aircraft by the enemy, in part from the heightened risks that pilots took to fulfill their missions.

Even if the Luftwaffe was better able to replace air crew and planes in the course of the campaign, it seems logical to presume that the Battle of France was a rude test for the German air force, and that it needed a pause to restore its strength before the Battle of Britain. The German archives will hopefully permit the verification of these hypotheses, and perhaps indicate to what degree the Battle of France weakened the Luftwaffe, helped delay the Battle of Britain, and perhaps indirectly helped the British to win it.

Other aspects of the struggle need to be better known, notably the battle with the Italian Air Force, which began when Fascist Italy took advantage of the destruction of French armies take part in the German victory. It would appear that just as Italian ground units could make little headway against the few French troops opposing them, the Regia Aeronautica also suffered heavy losses in the air battles, thus on June 15, in less than half an hour, Adjudant Le Gloan shot down five Italian planes.

Whatever the losses in the air battles over France from May 10 to the Armistice, they indicate very clearly that French airmen fought and fought well. No doubt the French infantry, constantly menaced by German bombs, had quite another opinion; they claimed they never saw a French plane over their heads. Their view, if mistaken, is nonetheless understandable. They did not know that the air battle was being fought at high altitude; even when French aircraft flew low, particularly to "show the colors" to the ground troops on what came close to being sacrifice missions, the ground troops did not recognize them. French aircraft were struck, and sometimes brought down, by the fire of ground units who could indeed claim in good faith that they had not "seen" French aircraft.

As for the high command of the Army, which had little faith in the air weapon, had little understanding of its proper use, and had consistently fought to limit the funds accorded it, the reproaches which it addressed at air leaders, at the aviation industry, and finally at the airmen themselves were at the very least unseemly. Behind these reproaches there may be something besides incomprehension: a determination to expose the errors of an institution which wanted to fly under its own power, to free itself from subjugation to the Army from which it had sprung. It is significant to note that both General Giraud in 1943 and General Juin in 1946 saw in aviation an army whose essential role was to support the land army; the struggle for air superiority meant nothing to them.

These same conclusions were drawn by the French high command from its study of the Battle of France, and they were not the only conclusions: The French Army had been surprised by the German use of tanks and of a terrifyingly effective weapon, the dive bomber, the famous JU-87 Stuka; hence we lost the war because we had not developed the dive bomber. Such nonsense was even subscribed to by a few airmen. Apparently the doctrine of the Luftwaffe was not sufficiently known, for it was not understood that the use of Stukas and Henschel observation planes was only possible for two reasons: Germany possessed sufficient mastery in the air to limit the activity of Allied fighters, and French ground units did not possess the anti-aircraft weapons necessary to deal with aircraft which were very vulnerable during their dives.

These ideas held by the Army and put forward at the Riom trials also masked some ulterior motives of political nature. Marshal Pétain, the new French chief of state, had had until 1936 considerable direct and indirect influence on French military policy. By making a sacrificial goat of the air arm, which had passed into the hands of "unworthy" ministers like Pierre Cot, one of the most hated men of the French left, it was possible to avoid any indictment of the Marshal and his entourage. This consideration explains the political dimension of a subject whose nature was essentially technical. In the assault on the air force and its political leaders, the attackers were really striking at the Popular Front, which was responsible for social legislation which its enemies felt made Frenchmen lose their work ethic and dedicate themselves to strike activity and leisure.

The debates at the Rion trials demonstrated the futility of such arguments, though on some those arguments were to make a lasting impression. From that point on, the debate would no longer be objective, and it would be further complicated by the conflict between Vichy France and Free France. Even today it is difficult to treat these questions without running the risk of a sort of excommunication. This risk is all the greater if the subject to be treated is the period from the Armistice of 1940 to the liberation of France.

THE ARMEE DE L'AIR OF THE ARMISTICE

At the time of the Armistice the bulk of French military aviation was in North Africa: thirteen fighter GROUPES, twenty-one bomber GROUPES, five reconnaissance GROUPES and twenty-two observation GROUPES. These GROUPES, comprising 851 airplanes, were not capable of military operations, in part because they lacked maintenance facilities and in part because they had been rendered incapable of flying on orders from the high command, this latter measure having been taken to prevent airmen from flying off to join the "dissidents" in Egypt or Gibraltar. It was in such conditions that the Armée de l'Air had to contend with Operation Catapult, directed against the French fleet at Mers-el-Kebir; only the fighters were able to get into action, shooting down two British planes.

Operation Catapult did have some beneficial results. The Armistice agreements had called for the dissolution of air units. After Mers-el-Kebir the German and Italian Armistice Commissions decided to accept the maintenance of a certain number of units to assure air protection of French territory in the event of British "aggression." The air forces permitted by this accord comprised a headquarters located in France which controlled two air regions (the first in Aix-en-Provence, the second at Toulouse), a superior air command in North Africa encompassing air commands in Tunisia, Algeria and Morocco, and five autonomous air commands in the Near East, West Africa, Equatorial Africa, the Indian Ocean, and Indochina. These various commands could muster: in France six single-seat fighter GROUPES, two two-seat fighter GROUPES, four bombardment GROUPES, two ground attack GROUPES and three reconnaissance GROUPES; in North Africa six single-seat fighter GROUPES, nine bomber GROUPES and six reconnaissance GROUPES; in the Near East one fighter GROUPES, one bombardment GROUPE, one GROUPE of six escadrilles of reconnaissance craft (Potez 25); in West Africa a fighter GROUPE, a fighter escadrille, three bombardment GROUPES, and two reconnaissance escadrilles. Equatorial Africa, which would soon join Free France had two mixes escadrilles, while Madagascar had only one mixed escadrille. And finally, in Indochina there were perhaps a hundred aircraft and a few flying boats, the aircraft ranging from the Morane 406 to the Potez 25.

At this juncture occurred the Dakar affair. A British fleet, accompanied by Free French, including General de Gaulle, appeared off Dakar in an attempt to rally French West Africa to their cause. But when Governor General Boisson rejected General de Gaulle's proposals, hostilities began. There was a naval bombardment and aircraft from a British carrier opposed Curtis M-75 fighters, while French Glenn Martins bombed the British fleet, scoring a direct hit on a cruiser. In all one Curtis was lost and five British aircraft were destroyed. The affair ended there, for the British cancelled the operation.

As reprisal for the attack on Dakar, General Bergeret, the new Air Minister, decided to stage a bombing raid on Gibraltar. On September 24, 1940 an expedition of some sixty aircraft dropped forty tons of bombs on the strategic base, and the following day eighty bombers renewed the attack, dropping fifty-six tons of bombs on the British port. The results were not very satisfactory, for large numbers of bombs missed their objectives. Was this the result of human error, the lack of good bombsights, or simply the result of a lack of enthusiasm for bombing yesterday's ally? It would be interesting to know the interpretations given to these raids and their observed result. It is not at all certain that French air crews, still smarting from the attack against Dakar, and especially the attack on Mers-el-Kebir, would have been inclined to spare the British. It would be apparent later that they were not so inclined in Syria, and at the time of the Allied landing in North Africa.

But other battles awaited the Armée de l'Air of the Armistice -- in Indochina, in Syria and in Madagascar -- before the Allied invasion of North Africa produced fundamental changes. In Indochina there was a war between Siam and the French forces in the area, a war not known to the French people at the time, and one completely forgotten today. This war began on November 25, 1940 with a Siamese attack and ended on January 28, 1941, thanks to Japanese mediation. French air forces were engaged and the French fighter force, equipped with Morane 406s, won a number of victories despite their inferior numbers, and despite aid which the Japanese gave to the Siamese in the form of aircraft and probably of pilots. This would be an excellent subject to explore further, especially for researchers with access to Japanese sources.

The Syrian affair was an even more serious confrontation. Once again French forces opposed British and Free French units. The British found a pretext for the war in the use ot Syrian and Lebeanese airfields by German aircraft transporting supplies to Iraq, where a pro-Nazi uprising had broken out. Beginning on May 15, 1941 British aircraft bombed French airfields in the Levant. These attacks led the Germans to authorize the sending of reinforcements in the form of fighter GROUPE III/6, equipped with the Dewoitine 520. On June 8, British ground forces crossed into French held territory. The French authorities threw what aircraft they had into the battle, and sizable reinforcements were sent to the area by way of Italy and Greece: four bombing GROUPES equipped with LéO 45s, two Aéronavale escadrilles equipped with Glenn Martins, and a fighter GROUPE with Dewoitine 520s. At first these forces were able to hold mastery of the air and assist in the ground fighting. The introduction of Curtiss P-40s by the British changed this situation, and on July 8 General Dentz asked for a cease-fire and on July 14 an armistice was signed at Saint-Jean d'Acre.

The air war in Syria is an interesting one to study because it throws into high relief both the strong points and the weaknesses of French aviation of the Armistice era. The aircraft flown were more and more outmoded -- the Dewoitine 520, for example, when faced by the P-40. Then too, the French units were not able to maintain their fighting trim or their aircraft, due to restrictions imposed by the Armistice Commissions. Then too, flying personnel whose hours of flying time had been sharply cut lost that sharp edge of proficiency, though successes against the British somewhat masked this shortcoming. For these reasons alone, the study of the Franco-British air war in Syria and Lebanon is a worthwhile undertaking, though many other aspects of the conflict could also bear inquiry.

In Syria we encounter once again the matter of motivation among flight crews who were sent out to do battle with those who had until recently been allies. There is no doubt that for a time after Mers-el-Kebir and Dakar there was a certain Anglophobia in the French military. But for the majority of airmen it was not sufficiently strong to obscure the fact that their first enemy remained the Germans. Yet the air battles were bitterly fought, without evidence of the slightest concession being made to the adversary. Later, after the Armistice of Saint-Jean d'Acre, when the former combatants were given the opportunity of joining the Free French, very few took the opportunity.

The reasons for the failure to rally to General de Gaulle are not easy to understand. Respect for discipline, fidelity to Marshal Pétain, Anglophobia, and anti-Gaullism do not explain everything, nor is it clear that they were mutually supporting motives. Did the sense of discipline and fidelity to the Marshal automatically rule out any sympathy for the Gaullish cause? This seems doubtful at the very least. For a considerable length of time, and despite demands from both Vichy and the Gaullists, many Frenchmen were persuaded that "Vichy" and "London" each was playing its allotted role in the service of France. Many of the military held this view, in Syria as well as in France and French North Africa.

The Anglophobia which was so widespread in the French Navy did not have very deep roots in the Armée de l'Air; yet it is possible that it developed there after the Armistice of Saint-Jean d'Acre because of the British attitude towards "Vichy" officers. Might this attitude have been adopted to discourage enlistments in the Free French forces? Did Free French who encountered "Vichy" officers after the fighting was over put forward serious efforts to recruit them? Were they all desirous of seeing brilliant officers join the Gaullist air units, where they could pose problems of competition?

Other factors no doubt help explain the small number of airmen who passed over to the Gaullist side, and among them there were family considerations. All of these factors are difficult to isolate because they were masked by a certain amount of moralizing which stressed considerations of honor and duty. There is no doubt that a sense of duty and a desire to keep one's word played a role in this affair -- but just where such considerations were a major factor and just where they were an easy alibi is not easy to tell. We might note that a number of airmen who had fought the British in Syria escaped from France after November, 1942 and went to North Africa or to England. But by November,

1942 events had changed many men's way of thinking, as we shall see.

An indirect result of the Syrian affair was the increased activity in military aviation that was permitted by the Germans. We will not dwell on this activity in detail, but only note that it merits investigation. In July, 1941 France and Germany agreed on a common aircraft construction program which was to produce by April, 1943 more than 3,000 aircraft and about 8000 motors, two-thirds going to the Germans and the rest to France. Later the French aircraft industry was to produce other planes for the Germans. In this activity there was certainly outright collaboration on the part of some industry officials, but there was also a not inconsiderable resistance, which took the form of clever and systematic sabotage of materiel destined for the Luftwaffe and for the air forces of its satellites. The operations of this industry and the research carried out clandestinely no doubt made it easier to resume aircraft production again after the war. It would be interesting to know more precisely the role of this activity in the period between the Armistice of 1940 and the renovation of the industry in the 1950s.

MADAGASCAR

In May, 1942 the British launched an expedition against Diego Suarez. The fighting lasted two days. Fighting broke out again on September 18 and did not end until November 6, when Madagascar had been completely occupied. French air strength on the island was extremely limited, yet the airmen fought with determination. The personnel taken prisoner by the British could not return to Vichy-held territory and thus they became the first who became part of the Free France through "geography." Some of them would later fight in Free French units. Here again it would be interesting to know their feelings toward the Free French and toward the British. For various reasons it appears that they faced different problems from those in Syria or North Africa.

NORTH AFRICA

Two days after the end of hostilities in Madagascar, American and British troops landed in Algeria and Morocco. Once again, following orders given by the hierarchy in place, the men of the Armée de l'Air threw themselves into a fratricidal battle which they could not hope to win. The air battles were particularly hard fought over Oran, where

French fighters brought down seventeen Allied planes, and in Morocco, where fighter GROUPE II/5 brought ten American planes at the cost of five pilots killed, including the GROUPE commander. It is sad to relate that in these battles GROUPE II/5 -- the "Lafayette" GROUPE which had inherited the traditions of the escadrille in which so many Americans had fought in the previous war -- lost in one morning more pilots than it had lost in the Battle of France. On November 10 the fighting came to a halt and the presence of Admiral Darlan at Algiers helped bring the return of peace and order.

In France units were placed on alert but were not sent into combat against the Germans, who, in violation of agreements, invaded unoccupied France on November 11, 1942. Thus the armed forces, which had been made to fight against our former allies out of a sense of honor, now were kept from engaging the invading German troops. Several arguments have been advanced to justify this non-intervention, particularly by the losses among flying personnel that would have occurred. These risks were not evoked, however, when it was a question of fighting the Allies. Didn't honor require that French troops fight against the German invader? A number of officers and non-commissioned officers decided that when the Marshal refused to give them the order to fight, he broke the bond that he had made with them when he had them take an oath to serve him for the honor of French arms. Since honor had not been defended, the oath no longer bound them.

We might also ask why the order was not given for air units to fly to North Africa. Such an order would have been easy for bomber and reconnaissance units, since their aircraft had sufficient range to cross the Mediterranean. For fighters the problem was not so easily resolved, but a stop in Corsica was possible and the risk could have been run.

After November 11 everything was changed. On November 27 the Germans and Italians decided to dissolve the French forces permitted under the Armistice. Numerous personnel received what was called "armistice leave." In this case they continued to receive a portion of their pay and hence had a certain liberty of action. Some took civilian jobs, others joined the Free French forces or took part in the resistance. Yet the Armée de l'Air of the Armistice did not completely disappear, since in March, 1943 a General Secretariat of Air Defense was created, with an air defense staff and three subsidiary INSPECTORATS, one for anti-aircraft defense, one for air security, and one for civil defense. Was this a case of putting Frenchmen to work for the German benefit, or was it a means of stocking personnel who could serve against the Axis when the opportunity presented itself? These contradictory theses are still being defended, and it is difficult for a French researcher to weigh them because of the emotional reactions they still provoke. Perhaps an American scholar, working as Robert Paxton did on Vichy France, could arrive at balanced judgements accepted by those on both sides of the issue.

Similar judgements are needed in the matter of "Jeunesse et Montagee" (Youth and Mountains). In order to perpetuate the concept of military service, the Vichy government had created "Chantiers de Jeunesse" (Youth Workshops). Along with these workshops the Armée de l'Air had provided the cadres for "Jeunesse et Montague", a civilian organization which it controlled, and this organization which continued until 1944, needs to be studied. Was this a means of preparing cadres for the resistance, or for protecting young men, especially those who had come from the Air Academy and risked being sent to Germany as part of the compulsory labor service, or yet again something else? Was the civic and moral training given these troops in the unswerving lines of military tradition or is that tradition adulterated by the "Vichy spirit" a mixture of Boy Scoutism and authoritarianism well symbolized by the song "Maréchal nous voilà," and by an ill-defined ideology expressed in the formula "travail, famille, patrie" (labor, family, fatherland) and the exaltation of peasant virtues and a return to the soil. An objectively pursued analysis should provide important information for the political scientist, for this "maréchaliste" spirit reappears in French political history, and paradoxically under the Fifth Republic.

THE FREE FRENCH AIR FORCES (F.A.F.L.)

During the period before November, 1942 there were those who continued to fight against Germany in the Free French Air Forces. Much has been written about the operations in which these French airmen took part, either in French units created within the structure of the R.A.F., or in R.A.F. units; consequently we need not dwell on that subject here. We should note however that F.A.F.L. units were active in many different places, in Great Britain, in Greece, in Cyrenaica, in the Middle East, in Ethiopia, and in the Chad, where they supported the operations of Leclerc's columns.

If the battles of these men are well known, they themselves are not. Sociologically speaking, who were they? What was their age, their military and their aviation backgrounds? What were their motivations, and by what means had they joined? How did the new arrivals fit into the older units?

How did Frenchmen who fought in the R.A.F. see their British comrades in arms? How did they accept a different way of doing things? As for those who preferred to serve in all-French units within the R.A.F., what were the reasons for their preference? How did they feel toward the R.A.F. hierarchy above them, and how did they react to the criticism levelled at them by other Frenchmen, men who hesitated to fight themselves, but not to criticize those who did. What was life like within the units, and what was morale like? We assume today that morale was good, since they were on the winning side, but was it good in those dark days of early 1941, when England stood alone against the German onslaught? How did the men react to events elsewhere, to Mers-el-Kebir, to Dakar, to Syria and to the landings in North Africa, and what were their feelings during that troubled period from November, 1942 to the summer of 1943? Until December 24, 1942, date of the assassination of Admiral Darlan, in North Africa they were viewed as traitors by an administration installed by Vichy and supported by the Americans. Anti-Jewish laws continued to be applied in North Africa and Gaullist sympathizers were imprisoned there. The emergence of Giraud did nothing to resolve this situation, and it was not until June, 1943 and the creation of the French Liberation Committee that the situation improved and the Free French there could be confident of their future.

Just as the R.A.F. could be studied through the perspective of Frenchmen who served in it, it would also be interesting to examine the Free French Air Forces as they were perceived by the R.A.F. What did the British think of these airmen, many of whom lacked experience? How did they receive them? Why was it that they accorded them relatively good billets within British units and even positions of leadership in RAF squadrons.

We may also ask how the British reacted to the decision to create a French air unit in Russia, the GROUPE NORMANDIE, later the Normandie-Niemen Regiment. This unit, created to support a policy, was later the subject of considerable writing, and became very well known. At the outset it was a Free French unit, but it proved impossible to compensate for losses by recruiting among Free French and it became necessary to send pilots serving in North Africa. How were such reinforcements received and what were the consequences on unit morale? This same personnel policy was also applied to other GROUPES; Bretagne, re-equipped with Marauders, was the most affected, but from 1944 on it also affected the F.A.F.L. GROUPES Lorraine, Alsace, and Ile-de-France.

While these changes were taking place air crews from North Africa arrived in England to be formed into a heavy bomber ESCADRE flying Halifax aircraft, as well as two fighter GROUPES to be equipped with Spitfires, one to fight as part of the Second Tactical Air Force, the other to become part of Air Defense. What were the relations between the airmen of Free France and those coming from other combat units? Were there not problems arising from difference in status or in points of view? Were the airmen from North Africa victims of some sort of complex? Since they had for the most part simply rallied the cause when their territory changed its allegience, did they not begrudge the Free French their achievements which had begun with an act of indiscipline.

To return to the airmen of Free France, it would be a good idea, once there were studies telling us who they were and where they came from, to find out what became of them afterward, whether in the service or in civilian life. Was their Free French label a help to them or a hindrance? Such inquiries would not be easy, since we are dealing with fairly small groups of men. In September, 1941 the "Gaullist" flight personnel (not including parachutists, who remained attached to the air force until 1946) numbered only sixty-nine officers (fifty-two pilots and seventeen observers) and 136 non-commissioned officers, of whom 134 were pilots. As of July 1, 1943, date after which no further enlistments were accepted, the flying personnel had only slightly increased.

THE REBIRTH OF THE ARMEE DE L'AIR 1943-1945

Once the bombing in North Africa succeeded, the French armed forces returned to the fight. Negotiations began with the Allies to determine how to rearm those forces, including the air force. As soon as they arrived, the Americans re-equipped, at least symbolically, the Lafayette Fighter GROUPE II/5. This GROUPE fought in the Tunisian campaign as part of an American force. Reconnaissance GROUPE II/33 and bombardment GROUPES I/25 and II/23 also participated in the campaign, using French materiel; moreover Fighter GROUPE

II/7, equipped with Spitfires, took part in the last part of the campaign.

It was not any distrust felt toward French pilots that made rearmament a slow process, but rather a series of unforeseen difficulties. Though mobilization had brought more manpower, the training of these men required a master plan of aerial rearmament, which would include training in American schools, or in French schools set up in Morocco beginning in January, 1943; as for crews already constituted, but who would require re-training, these could be formed into GROUPES, but the Allies required that those GROUPES be integrated into larger Allied air units. General Giraud had quite a different idea. His views on airplanes were about as clear and valid as the ideas on tanks which he gave General de Gaulle in 1938. To his way of thinking the air force should be at the service of the army, with French air forces placed under the orders of the ground commander. This was a return to the scattering into penny packets which had cost us so dearly. Fortunately the Allied command refused to accept such a solution; it ran the air forces as independent, the linkage with land forces being made at the army group level.

General Bouscat was able to convince the French ground commanders, not that the Allied views were the right ones, but that unless those views were adopted there would be no rearming of the French air forces. On July 1, 1943 General Bouscat became Chief of Staff of the French Air Forces and took over command of all aviation, including the Free French Air Forces. He quickly drew up a rearmament plan, Plan VII, which was accepted by both the Allies and the French government.

Without going into detail, we might note the nature and disposition of French air units in September, 1944, after the landing in southern France:

In England two heavy bombardment GROUPES formed into an ESCADRE and flying Halifaxes with Bomber Command, three fighter GROUPES operating with 84 Group, Second Tactical Air Force, and one GROUPE attached to Air Defense; then the Lorraine GROUPE flying Boston light bombers as part of 2 Group, Second Tactical Air Force.

In southern France three fighter GROUPES equipped with Spitfires and six groups equipped with American P-39 Aircobras and P-40 Tomahawks, later exchanged for P-47s; two reconnaissance GROUPES, I/33 with P-38 Lightnings and P-51 Mustangs, and II/33 with Spitfires and P-51s; finally six medium bomber GROUPES flying B-26 Marauders, formed into two ESCADRES under the 11th Bombardment Brigade.

In Russia, the Normandie-Niemen, now transformed into a regiment, and to which there were plans to add a ground-attack unit flying Stormoviks.

After the return to France the Atlantic air forces came into existence, flying German and French aircraft and operating over the German pockets of resistance at Saint-Nazaire, Royan and Lorient. More important was the creation of the First French Air Corps, under General Gérardot and including the First and Fourth Fighter ESCADRES as well as Reconnaissance GROUPE II/33. This body was placed under the First U.S. Tactical Air Force, which determined the use of air forces with General Devers' Sixth Army Group. But quite often the French units operated for the benefit of the French First Army of General de Lattre de Tassigny, sometimes with heavy losses. The General no doubt had difficulty accepting the fact that he could not issue orders to the French First Air Corps; this at least was the only explanation that could be found by airmen when they discovered to their surprise that "le roi Jean" (the name often given to de Lattre) had failed to mention them in his history of the French First Army -- this though they had fought alongside his forces through the campaign in Alsace and Germany.

This sort of attitude reveals something of the difficulties that high ranking Army officers would have after the war in understanding the rules for the use of air forces. The command responsibilities of the French Army's leaders did not go beyond the army level, Marshal Juin commanding an army corps and Marshal Leclerc an armored division. Since the use of air forces was handled at the army group level, neither those generals nor those on their staffs ever had occasion to work with air forces in that way. This may well explain some of the attitudes they were to hold in later years.

In May, 1945, at the time Germany capitulated, the French air forces could be proud of the battles they had fought to end the war and liberate France. But these forces, which had been brought into existence with Allied help, needed not only to be maintained but also renovated, if their fortune was to be assured in the age of jet power and the atom.

THE ARMEE DE L'AIR SINCE 1945

In 1945 the fundamental tasks facing the Armée de l'Air were several: it needed to put its house in order, to define the doctrine governing its role and the missions necessary to fulfill that role; it needed to prepare for the future and at the same time be ready to carry out the actions the government might assign to it.

The problems were far easier to identify than to solve -- and, for that matter such problems are not easy to understand today. One needs to recall, for example, the lack of homogeneity among the personnel then in service, who fell into a number of groups: F.A.F.L. personnel, former Vichyists, those who had been in the resistance, and those who had been taken prisoner in 1940. There was a division between those who had fought in some sort of unit after 1940 and those who had not. Among those who had fought it was necessary to distinguish between those who were in the R.A.F. and those who were with the Americans -- not to mention those of the famous Normandie-Niemen, who came back from Russia. In this connection an effort should be made to understand the influence of the Communist ministers, Fernand Grenier and Charles Tillon, who were at the head of the Air Ministry from April, 1944 to December, 1945. Their administrations sparked a quantity of polemics, but were they based on fact? It would also be interesting to study the internal workings of the administration. There was, for example, the general disorder that enabled certain officers -- or at least those who passed for officers -- to carry out some amazing commercial and financial operations which were in fact so many cases of swindling. There was also a lack of authority in Paris that led certain general officers ensconced in their proconsulates to accept from the central administration only those orders that suited them.

To re-establish and assure order it was necessary at the outset to revamp the central administration, and first of all the General Staff. An examination of the circumstances in which this task was accomplished and the time that was required should provide some insight into how well this work was done. We must begin with the men responsible for this work, first of all the ministers and chiefs of staff of the period. We have suggested earlier a study of the work of the Communist ministers. This approach could be extended by inquiring into the careers of their successors, who were secretaries of state rather than ministers. Firmly anchored in political affairs, they seem to have left much of their work to the chiefs of staff, who enjoyed a somewhat more flexible relationship with the ministerial level. To understand the actions of the first chiefs of staff in the immediate postwar period, the researcher would need first of all to know something of their careers during the war and in the prewar era. In a brief period of time there were five Chiefs of Staff of the Armée de l'Air: General Valin to February, 1946; then General Bouscat, who had given up his wartime title as Commander in Chief of Air Forces, until September, 1946; then General Gérardot to February, 1947; and finally General Lecheres, who would keep the post until August, 1953.

The importance of these postwar Chiefs of Staff, great as it was, should not lead us to neglect officers of lesser rank who held key positions in the General Staff: the MAJOR GENERAL, their immediate subordinates, and finally the bureau chiefs. We need to know who these men were, where they came from and what their previous military experience had been.

Among the tasks accomplished in the era was the pruning and the revitalization of the cadres, either through voluntary departures or through purging measures, these latter measures limited to fewer than three hundred cases. Then there was the problem of the schools. Should the Ecole de l'Air be reopened? The debate that arose over this question was a fascinating one which came to a happy conclusion: the Ecole de l'Air reopened its doors under conditions comparable to those obtaining before the war. The Ecole Supérieure de Guerre Aérienne also took a new lease on life under the designation Centre d'Enseignement Supérieur Aérien, which General Gérardot strove to make the antithesis of what he had known at the Ecole de Guerre of the Army in 1935 -- a school which he described as a "primary school for applied tactics."

This same period saw the creation of the REVUE DES FORCES AERIENNES FRANCAISES, in whose pages young officers found a place to shine intellectually in the fullest freedom. This journal is now a valuable source for historians, and we can only regret its disappearance in 1970 -- on the order of a Minister of Defense of extreme centralizing tendencies who was suspicious of any kind of opposition -- however intelligent, just, and well expressed it might be.

The General Staff had also the task of defining a new organization for the Armée de l'Air, which still rested upon the law of 1934 -- a law which had proved ill-suited to air operations. In 1945 it had been necessary to violate the law in order to create the Groupement des Moyens Militaires de Transports Aériens, which was charged first of all with returning to France deportees and prisoners of war, then with assuring transportation links between France and her empire. Illegal though it was the G.M.M.T.A. remained in existence until 1963, when it was replaced by C.O.T.A.M. (Commandement du Transport Aerien Militaire); this creation was made possible by the adoption of legislative and regulatory texts which accorded to the Armée de l'Air an organization suited to its mission. We will limit our treatment of this whole problem of organization, but an examination of the subject in depth, including the ways in which it was treated between 1945 and 1960, and the polemics and the legal difficulties it engendered, would show very well all the problems the Armée de l'Air had in getting rid of its original ground-oriented structure.

These problems were even more evident when it came to adopting a doctrine of use for the air forces. True to its prewar position, and fortified by the Allied experience, the Armée de l'Air held to the doctrine of securing air superiority before undertaking any other air action or any effort on behalf of land or naval forces. The provisional instruction on the use of air forces incorporating this idea appeared without the approval of the General Staff of the Armed Forces and in total contradiction to the views of that body. General Juin, who headed the General Staff, and who by virtue of that position in principle exercised authority over the Air Staff, saw the air arm only as a means placed at the disposal of the land forces. The arguments advanced along these lines -- arguments which sprang from the prewar ideas which had cost us so heavily in 1940 -- deserve serious study, for they reflect a basic military conservatism which can still be encountered today.

This quarrel was of no great importance save in the theoretical realm, for in any case the Armée de l'Air did not have the means to carry out the various types of missions that could be foreseen. Equipped with aircraft which had been given by the Allies and with German models whose construction was continued in French factories, the Armée de l'Air was all the more limited in its possibilities for operations because its technical and material bases were defective. The radios were outmoded, and radar not yet in place; the standard of living for personnel was deplorable, at least for those stationed in France. Despite effective assistance from Great Britain, it would take much time and American financial assistance to develop a modern infrastructure.

As for aircraft production, the aeronautical industry was still capable of producing airplanes despite the damage suffered during the war. But what should the factories produce? Should they continue producing German models, recommence the building of French models of 1940, or set their workers to making kitchenware? All of these ideas were considered and tried in conditions that were far from satisfactory. But then how could very good results be expected, considering the slender allocations given to the air arm? These grew less rapidly than the global budget and the general military budget. If comparison with the global budget indicates the priority given to reconstructing the country, a comparison with the budgets of the other services would lead to some interesting conclusions, as well as some instructive comparisons with other periods, including the most recent ones. A careful analysis of the various expenses for the air arm would give a fairly accurate portrait of the conditions of military aviation in that era.

Concerning the French aviation industry the Service Historique possesses the collection of Ingénieur Général Thouvenot, which demonstrates very well the problems encountered between 1945 and 1950. Once again it seemed wiser to prepare the future than to sacrifice it for a mediocre present. As in the early thirties, a policy stressing prototypes was developed, and though the policy met with strong criticism, it enabled design departments to make up much of the time they had lost during the war. There were some patent failures, like the Cormoran transport and the Arsenal VB10 fighter, and many prototypes were not followed by serial production, but the new designs proclaimed the renaissance of the French aeronautics industry. We might cite among others the SO 3000 Triton, which was the first French Jet aircraft to fly on November 14, 1946, the SO 6020 Espadon, the Nord 1405 Gerfaut, the Leduc 022, the Nord 1500 Griffon and above all the M.D. 450 Ouragan, ordered for serial production to equip the Armée de l'Air. The Ouragan was the first of a line that would lead to the Marcel Dassault Mystères and Mirages. Such were the results as alongside nationalized firms private enterprises were reborn: Marcel Dassault, mentioned above, Breguet, Potez, etc.

It would be interesting to study the relations of these firms with one another and with the government, not only for this period, but from 1945 to the present. Relations with the government would need to be studied at all echelons, from the ministerial level to that of the Direction Technique et Industrielle de l'Air, and on down to the engineers of the execution level. In making such a study one would have to keep in mind the various ties and relationships in this military-industrial ensemble, the backgrounds of the men who made it up, most of them coming from the Ecole Polytechnique and the Ecole Supérieure de l'Aéronautique. Collaterally, there should be a study for the entire period of the aircraft motor industry and the industries supplying various other components.

As both participant and frequent beneficiary in this military-industrial complex, the Armée de l'Air continued to be faced with sometimes contradictory obligations. Until 1950 it was charged with fulfilling with very limited means three separate missions: the restoration of its strength within France, liaisons with the Empire, sometimes punctuated by interventions, as in Madagascar, and finally the war in Indochina.

The Indochina War merits considerable attention, first of all because the operations there are full of lessons, particularly concerning the relations between army and air force commands. Now that these archives can be consulted, it is tempting to form hypotheses on the responsibility for certain military failures, notably Dien Bien Phu. These hypotheses will no doubt yield conclusions at variance with those drawn by General Navarre or by Monsieur Pleven, then Minister of War. There is another point which needs to be examined: during the Indochina War and after the Geneva agreements much emphasis was placed on the difference in viewpoint between officials in France and the professional soldiers who were fighting in Indochina to hold on to this territory. The loss of the war has been laid at the door of civilian leaders, but in fact the war was also distasteful to a number of military chiefs. It is interesting to note the lack of haste with which the successive chiefs of staff of the Armée de l'Air committed their resources to the conflict. For them the essential concern was to fulfil in Europe the promises we had made within the framework of NATO. Something of the same tendency can be seen in more muted form during the war in Algeria. This conflict will also offer numerous subjects for dissertations when the archives concerning it are opened. By the same token, N.A.T.O. offers rich fields of study, including the circumstances of its creation, the viewpoints of the Armée de l'Air faced with the missions which fell to it, and the technical and financial relations with our allies, and particularly with the Americans.

There are considerable contrasts between the 1950s and the 1960s which are of interest to the historian. From the time that he came to power, General de Gaulle resolved to reassert his country's political independence; to that end he decided upon a nuclear striking force, whose realization from 1963 to 1971 was exclusively the charge of the Armée de l'Air. Here again a study of the legislative and budgetary framework would offer a number of insights into the government's military policy, which on the surface seemed simple and straightforward, but was in fact full of contradictions. For the time being research must be limited to the materials so far made public. This is also the case with more recent periods where it is difficult to do research in any depth. Apparently there has existed within the country a consensus on the nation's military policy, which has scarcely changed from the time of De Gaulle up to the present. While there has been a general acceptance of basic principles expressed in the policy of dissuasion, one should remember that military policy, like war itself, is -- in the words of Napoleon -- "an art in which everything is in the execution." If French defense policy over the past years were examined through plans, budgets and the various political debates over it, such a study would alter somewhat our judgements on French defense policy. While at this point we reach the limits imposed by the nature of this present work on military aviation, there is a final question for which an answer must be sought: Do Frenchmen understand that within the confines of political option defined by succeeding governments since 1950, the use of French military power presupposes an air force with sufficient means to assure the freedom of action of other forces? Without those means in the event of a conflict France would find herself back where she was in May, 1940: condemned to lose the battle and then her freedom.

PART II

BIBLIOGRAPHY

GUIDE TO ABBREVIATIONS OF PERIODICALS

A.A.	AIR ACTUALITES
A.H.	AIRPOWER HISTORIAN/AEROSPACE HISTORIAN
F.A.F.	FORCES AERIENNES FRANCAISES
L.A.	LES AILES
M.A.	LE MONITEUR DE L'AERONAUTIQUE
R.A.A.	REVUE DE L'ARMEE DE L'AIR
R.A.M.	REVUE DE L'AERONAUTIQUE MILITAIRE
R.D.N.	REVUE DE DEFENSE NATIONALE
R.F.A.	REVUE DES FORCES AERIENNES
R.H.A.	REVUE HISTORIQUE DES ARMEES
R.H.M.C.	REVUE D'HISTOIRE MODERNE ET CONTEMPORAINE
R.M.A.	REVUE DU MINISTERE DE L'AIR

GUIDE TO RESEARCH MATERIALS

ARCHIVES OF THE ARMEE DE L'AIR

These are the most important archives for the history of French military aviation. They are divided into series, then into subseries and subdivisions. While detailed inventories may be consulted at the Service Historique de l'Armée de l'Air located in the Chateau of Vincennes, the series are listed here along with a brief description of their contents.

SERIES AA.

This series covers military aviation from its origins to 1914. It is unfortunately a relatively small collection. Research in this period requires the examination of a great variety of materials, with archival sources often no more than fragmentary.

SERIES A.

The subject of this series is military aeronautics from 1914 to 1918. The materials in this series were turned over to the Armée de l'Air upon its creation, having previously been held by the War Ministry. This explains why the documents it contains relate only to Army headquarters, the central administration and aeronautics at the level of the field armies and army groups. Archival materials relating to aeronautics at army corps and division level were considered inseparable from archives of the ground units; as a consequence these will be found at the Service Historique de l'Armée de Terre, also located in the Chateau de Vincennes. Although reduced in size by losses suffered in World War II, the archives for the period 1914-1918 are of great importance. Along with their purely historic interest they offer sometimes touching testimony to the human dimension of the war, as in the war diaries of units like Groupe de Bombardement No. 1 or Groupe de Chasse No. 12. The series also contains post-war studies of the period 1914-1918, among them an important critical study of the victories of French aces.

SERIES B.

Here are grouped records of the interwar period, notably records of staff and services. The collection is very incomplete for two reasons: first, since the Armée de l'Air did not gain full independence until 1933, many documents before that date were retained by the Army; then too, heavy losses to this collection were incurred in June, 1940, when the archives were being moved to the south of France by truck. Despite its incompleteness, this series has considerable interest for the historian. It provides information on the Deuxième Bureau (intelligence) and also on the disarmament conferences. From 1934 on the archives of the Conseil Supérieur de l'Air offer a valuable perspective, and from 1938 onwards the records are more complete, permitting a study of operational and industrial planning. This series also contains valuable data on purchases of aircraft and other materiel from the United States.

SERIES D.

Here are grouped the records relating to the Second World War, arranged in several subseries. For the period September 1939-June 1940, there are two distinct collections: first of all there is the Air Ministry collection, a continuation of the interwar series, and which concerns the general effort to obtain the necessary means for the war; then there is the collection of the Grand Quartier Général Aérien, or Air Force Headquarters, from September 3, 1939 to June 25, 1940, covering operations and logistical matters relating to the fighting units.

From the period June 1940 to May 1945 there are several different collections corresponding to various commands and phases of the period. The Vichy collection, which is the most numerous, contains the papers of the Secrétariat d'Etat de l'Aviation (1940 to November 1942) and of the Secrétariat Général de la Défense Aérienne (1943 to June 1944). The largest element in this collection is composed of the papers of the General staff of the Armée de l'Air, and from 1943 on the staff for air defense. The Vichy collection contains much interesting material on the Armée de l'Air during the Armistice and its internal operations, on relations with German and Italian Armistice Commissions, and on measures against Gaullist "dissidents."

The Vichy collection has three valuable groups of material on the Armée de l'Air before June, 1940. The first centers around the Riom Trials, of which the S.H.A.A. has a stenographic record, as well as materials submitted to the court regarding the Armée de l'Air. The most valuable of these is the Chossat Report, which offers the most accurate and exhaustive study of the Armée de l'Air between 1933 and 1940, including plans, budgets, problems of airplane production, etc. A second group of documents is composed of critiques of the campaign of 1940 prepared by a number of unit commanders, and a synthesis of these critiques prepared by Commission G in 1942. The Commission's report also provides insights into the campaign in the Levant from May to July 1941. A third and final element is composed of the HISTORIQUES or outline histories of each ot the fighting arms, specifically fighters, bombers, reconnaisance, observation, and aerostation. These HISTORIQUES are supplemented by very detailed accounts of the air campaign of May-June, 1940 and the functioning of the high command in that period. Here are also to be found details of projects for intervention abroad, notably in Finland, the Caucasus and Salonika. Also grouped with this collection is a synthesis of intelligence estimates on foreign air forces of the thirties.

The F.A.F.L. collection, covering the Free French Air Forces, contains materials from 1940 to 1945; though it covers one of the most important phases in the history of the Armée de l'Air, it is unfortunately one of the smallest collections. The basic reason is that for operational and logistical matters the F.A.F.L. was dependent upon the Allied Command, thus its own archives relate essentially to personnel and administration. The personnel materials are in a sense the other face of the Vichy documents, since Vichy "dissidents" became Free French airmen. Among these airmen were paratroops who were at the time part of the Armée de l'Air.

The history of the Free French Air Forces can also be approached through the archives of the various fighting units such as Lorraine, Alsace, Ile-de-France, Bretagne, Normandie-Niemen, and others. These materials are in Series G, which will be treated subsequently. Then too, there are materials in British archives which complement those in France, and offer at the same time different perceptions. Finally, the S.H.A.A. possesses a microfilm on the Normandie-Niemen Regiment furnished by the Soviet authorities, and which is chiefly composed of press clippings relating to the Regiment. It is quite possible that the Soviet Union has archives on the same subject, but this is not known at the present time.

A final collection in Series D is that of the Forces Aēriennes Françaises. In theory this collection is posterior to November 8, 1942, the date of the Allied landings in North Africa; in fact it contains the war diary of the Air Command for French North Africa beginning in 1940; the diary is of special interest for its account of Allied air operations at the time of the landings. Until July, 1943 this collection concerns essentially French air forces in North Africa and in French East Africa. Here are to be found plans for the rebuilding of the Armée de l'Air, for which American aid was sought, and also French air operations against Axis forces in Tunisia. After the fusion of the air forces dependant on Algiers and the F.A.F.L., which took place in July, 1943, this collection covers all units of the Armée de l'Air under the command of General Bouscat.

When the General Staff of the Armée de l'Air resumed its normal functions in Paris in October, 1944, it took over a part of the activities of the Algiers command (we will treat subsequently the question of these archives). The Command in Algiers kept the control of air units engaged against the enemy, so its collection contains the archives of these operational units, which were divided into three groups: those under American tactical command and contained in the Bombardment Brigade and the First French Air Corps; those under British operational control, under the designation Air Forces in Great Britain, even though they might be stationed elsewhere; and those under French command, essentially those under the Air Staff for the Atlantic Front.

SERIES E.

As noted above, the General Staff of the Armée de l'Air resumed its functions in October, 1944, and from that date down to the present its archives are grouped into Series E. The arrangement of the documents corresponds to the various branches and services placed under the Air Minister or the Chief of Staff. Here may also be found the archives of the Direction Technique et Industrielle de l'Air (D.T.I.A.), though D.T.I.A. documents are now passed to the Armament Archival Center of Chatellerault. Series E is open to the public up to 1958, in accordance with the thirty-year rule of French archives. There is one exception: archives of the Deuxieme Bureau after 1945 remain closed.

SERIES C.

This series is concerned with Indochina and contains some materials prior to 1945, including materials on the Franco-Siamese War, the Japanese occupation, and the coup of

March 9, 1945, along with military events after the coup. The post 1945 archives deal with the re-establishment of the air forces in Indochina, their logistical support and their operational use up to 1954. Here also are materials on the Geneva negotiations, the accords reached there, and their application. There is also data on the French air forces which remained in South Vietnam until 1956. One particularly interesting item is a general study on the operational use and function of French air forces in the Indochina War, along with the documents from which it was drawn. It should be noted that a part of the archives relating to Indochina will be found in Series E, cited earlier. With certain exceptions -- including Deuxième Bureau archives -- the materials in Series C are open to the public.

SERIES I.

This series relates to Algeria though once again relevant materials will also be found in Series E, especially concerning military operations against the F.L.N. In Series I will be found much material on the war, especially concerning the Fifth Air Region, Groupements Aériens Tactiques Nos. 1, 2, and 3, and the Groupement d'Aviation Légère. Unfortunately a part of archives were lost in the course of shipment to France in 1962, so the series contains many gaps.

SERIES F AND G.

Series F concerns air bases and Series G relates to units of the Armée de l'Air; the archives regarding both are turned over to the S.H.A.A. only when the base or unit ceases to exist. Part of these archives have no historical value, and part are included in other collections. The two series contain the daily records of both bases and units, and these records are of three types: first there are the war diaries maintained by units on campaign which chronicle the events and the activities of the airmen, with the official narrative sometimes leavened by more anecdotal items; then there are the unofficial records that many units kept at various times. These relate the everyday life of the unit, often in humoristic fashion. They sometimes contain veiled criticisms and are useful in gauging morale and opinion. In principle the S.H.A.A. has only the records for units that have been abolished; other records are still with the units and can be consulted with the approval of unit commanders.

A third type of record came into being in 1964, when the Armée de l'Air made obligatory the preparation of a record of activities and operations by each base and unit. These offer a complete profile of the unit for the period covered -- organization, personnel, materiel, missions, etc. These reports are produced every six months; moreover the report at the beginning of each year contains an annex regarding morale in the unit. These reports are in a sense historical in nature, and a capital source for the recent history of the Armée de l'Air.

SERIES Z.

This series comprises archives that are private or semi-private and that have been given to the S.H.A.A. or placed on deposit there, either by individuals or by institutions. These collections vary greatly both in their volume and their historical interest. Among the more important are: the Guy La Chambre Collection, papers of the Air Minister from 1938 to 1940; the Levy Collection concerning collaboration under Vichy; the papers of Ingénieur Général Thouvenot, concerning French aircraft production in the period to 1939, the purchase of American airplanes in that year, and restarting French aircraft production after 1945; the Gérardot Collection, concerning primarily the First French Air Corps commanded by General Gérardot, who subsequently became Chief of Staff of the Armée de l'Air; the Lauzin Collection, which contains interesting material on Dien Bien Phu; and the Brocard collection, particularly important for the air war, 1914-1918.

SERIES P.

Here are preserved the personnel folders of officers, these being confided to the S.H.A.A. one hundred years after an officer's birth. In principle these dossiers may not be communicated to the public until one hundred and twenty years after the birth of the officer concerned. However certain information regarding personnel may be communicated to researchers by special permission.

The dossiers of officers of more recent service, as well as some dossiers of non-commissioned officers, are kept at Chartres in the Bureau Central d'Incorporation et d'Archives de l'Armée de l'Air (B.C.I.A.A.A.) This agency, which is technically under the S.H.A.A., also stores purely administrative papers, though these may have historical interest. Certain of these, such as ledgers which relate to the administrative existence of a unit in wartime, may be consulted at Chartres, or at the S.H.A.A. with prior notice of one week.

ORAL HISTORY AT THE S.H.A.A.

Beginning in 1974, the S.H.A.A. undertook systematic interviewing of persons who had played direct roles in the history of military aviation. These have included a number of ministers and generals, as well as officers, pilots, observers, machine gunners, and mechanics who related in loosely structured interview their lives in the Armée de l'Air, their roles as leaders, subordinates or combattants, and their reflections on their careers. Preserved on tape, these interviews are a rich historical source, shedding new light on the written records. They are an essential source for studying the climate of opinion. To be sure they, like all such testimony, should be cross-checked with the historical record. Unless the interviewee has fixed special conditions for their use, these interviews may be utilized by researchers under essentially the same conditions as those established for written records.

THE S.H.A.A. PHOTOTHEQUE

The S.H.A.A. possesses large numbers of photographs, posters, and aerial maps useful in understanding the history of military aviation as well as in illustrating it. These holdings are available to researchers virtually without restriction, and copies may be furnished in accordance with regulations governing duplication. In addition to the photographic resources of the S.H.A.A., there are also collections at the Etablissement Cinematographiqũe et Photographique des Armées, at the Musée de l'Air, and for the period 1914-1918, at the Bibliothèque de Documentation Internationale Contemporaine.

THE SERVICE HISTORIQUE DE L'ARMEE DE TERRE

In the archives of the French Army will be found a number of documents on military aviation before 1914, as well as materials on aviation at various command levels during the war; this is especially true for army corps aviation. There are also important reports on aviation prepared for the Minister of War and other documents predating the creation of the Air Ministry. Even after that creation, the Army archives contain valuable materials on relations between the two services, including a number of reports and accounts of meetings of the Conseil Supérieur de la Guerre. Finally there are documents relating to French air policy up to the present.

THE SERVICE HISTORIQUE DE LA MARINE

This organism, also located in the Chateau of Vincennes, is the point of departure for researchers interested in the development of French naval aviation. It maintains an excellent library and also administers the archives of the Marine Nationale, though only a portion of those archives are located at Vincennes. Archival materials relating to naval aviation and aerostation have been arranged for use by researchers up to the year 1920, and work is currently under way to open materials for the interwar period. For the pre-1920 holdings here is a typescript guide available in the S.H.M. reading room entitled: "Archives Centrales de la Marine. Répertoire numérique détaillée." Aeronautical materials are in Subseries Gn, Service de l'aéronautique et des patrouilles aériennes. The subseries contains 190 cartons, of which the first fourteen relate to the period before World War I. The material for 1914-1918 is for the most part arranged by subject -- personnel, material, signals, etc.

THE ARCHIVES DU GENIE

The archives of the French Army's Corp of Engineers contain important materials for the development of military aerostation from the Revolution onward; they hold essential documents on the installation of Chalais-Meudon and the work which went on there, as well as information about the earliest military use of the airplane.

THE MUSEE DE L'AIR

Though the Musée de l'Air is not an archive in the normal sense, it does have a number of collections which it has received from private sources. Moreover it has numerous scientific and technical studies on aeronautics. Then too, the study of the early machines in the museum's holdings can provide answers to questions raised by historians concerned with technical evolution.

THE CENTRE D'ARCHIVES DE L'ARMAMENT DE CHATELLERAULT

This archival center was created in 1969 to receive the archives of the Délégation de l'Armament, and here are numerous materials concerning the Direction Technique des Constructions d'Avions (D.T.C.A.). In addition, it has become more and more common for aircraft manufacturers to confide their archives to this center, which is thus becoming a major repository for historical materials of technical and industrial nature.

THE ARCHIVES DE FRANCE

Many archival sources concerning aviation are to be found in the Archives Nationales and other public repositories. These sources may be political in nature, as in the Dautry Collection, given by the Minister of Armament of 1939-1940; they may be economic, as in the case of the records of the Ministry of Finance; or Judicial, as in the case of papers of the Haute Cour de Justice, which are not for the present available to researchers. It should be noted that the records of the Ministry of Colonies are a rich source, for that ministry controlled aviation in the colonies for a considerable period of time. Finally, we should note that both departmental and communal archives would contain materials on air installations within their confines.

THE ARCHIVES PARLEMENTAIRES

The archives of legislative committees are extremely rich and shed a revealing light on parliamentary debates. The number of committees interested in aeronautics is extensive, and it is not possible to offer a complete list here; their names changed from time to time, but in general they were concerned with foreign affairs, finance, or military matters. By way of example, for the Chamber of Deputies in the interwar period valuable information would be found in the papers of the Commission des Affaires Etrangères, the Commission des Finances, the Commission de l'Armée, and the Commission de l'Aéronautique (after 1928). Various other agencies could also have relevant archival material, among them the Foundation Nationale des Sciences Politiques and the Conservatoire National des Arts et Métiers.

MISCELLANEOUS DOCUMENTATION

This is a vast and varied mass whose limits are difficult to define. At the S.H.A.A. there are four distinct types of materials of possible use to the historian. There are first of all outline histories of the various units prepared by S.H.A.A. personnel; these histories are based upon archival sources and information supplied by former members of the units; these histories attempt to relate the life of the unit from its creation. There are also biographical files on prominent figures in aeronautics. These files contain a career outline, press clippings, and notes on bibliographic and photographic sources, and they are kept up to date. Then, there is considerable technical documentation in the form of various manuals and notices on utilization and maintenance of aircraft, etc. The S.H.A.A. also has complete collections of two important government documents: the JOURNAL OFFICIEL DE LA REPUBLIQUE FRANCAISE and the BULLETIN OFFICIEL DES ARMEES. Also useful to researchers is a collection of the JOURNAL DES DEBATS of both the Chamber of Deputies and the Senate from 1888 to the present.

ARCHIVAL SOURCES IN GREAT BRITAIN AND THE UNITED STATES

An air force is not just a means of warfare; it can also be a means of provocation or of dissuasion, hence it is important to know how it was perceived in other countries, and what its strengths and weaknesses appeared to be. The image of the Armée de l'Air can be seen in the files of other governments, where those files have survived and are open to researchers. This is true for both the United States and Great Britain. American military attachés filed numerous reports on French military aviation virtually from its birth and these have been preserved in Record Group 165, in the National Archives in Washington. By the same token French military aviation can be seen in the files of the Air Ministry, held at the Public Record Office, Kew; thus file Air 10/1648 contains a two hundred page "secret" report with three large appendices on the state of the Armée de l'Air in January, 1939.

As was noted by General Christienne earlier, the operational history of certain French air units attached to British or American commands can only be researched in the archives of those commands. For the period of the First World War much information on the French air service, its materiel, its schools, and other installations can be found in the Gorrell Collection, in Record Group 120, National Archives. The collection takes its name from Colonel Edgar S. Gorrell, who amassed and brought back to the United States a large quantity of documents on the air war for the preparation of a history of the U.S. Air Service. Also valuable for the period 1914-1918 are the interviews with American airmen who served with the French; these are held by the U.S. Air Force Historical Research Center, Maxwell A.F.B., Alabama. There is also a valuable collection relating to aeronautics in France in the Manuscript Division of the Library of Congress: the Tissandier Collection, begun by the French aeronauts Albert and Guston Tissandier, contains some 6000 items relating to aeronautics from 1776 to 1914.

BIBLIOGRAPHIES OF PUBLISHED MATERIALS

The most complete and reliable bibliography of French history is the annual BIBLIOGRAPHIE (6) published under the auspices of the Centre National de Recherche Scientifique. It indexes both books and articles from 1953 and has a special rubric "Aviation Militaire." A useful general bibliography of aerial warfare is the BIBLIOGRAPHIE (20) prepared by Karl Köhler of the West German government's Militärgeschichtliches Forschungsamt covering the most important works appearing through 1960. An excellent source for earlier literature, both books and articles, is the BIBLIOGRAPHY OF AERONAUTICS (8) which Paul Brockett prepared for the National Advisory Committee on Aeronautics from 1909 to 1932. For the period 1909-1916 alone this source lists nearly four hundred items under "France." A more recent American source is the INDEX (1) published by the Air University Library since 1949. It lists significant articles published in some seventy English-language military and aeronautical journals.

BASIC WORKS ON FRENCH HISTORY

Since the first part of the history of French military aviation came under the Third Republic, four basic works on that period should be singled out: Jacques Chastenet's multi-volume HISTOIRE (11), George Bonnefous' HISTOIRE POLITIQUE (7), François Goguel's LA POLITIQUE DES PARTIS (17), and volumes 10-14 of the series LA NOUVELLE HISTOIRE DE LA FRANCE CONTEMPORAINE, those by Mayeur (22), Rebérioux (26), Bernard (5), Dubief (15), Azéma (3), and Rioux (29). For the more modern period the researcher should take as point of departure the relevant volumes of the annual ANNEE POLITIQUE. For international relations there is the excellent HISTOIRE DIPLOMATIQUE (16) of Jean-Baptiste Duroselle, and the later volumes in the collection HISTOIRE DES RELATIONS INTERNATIONALES, directed by Pierre Renouvin (27, 28). For economic history Alfred Sauvy's HISTOIRE ECONOMIQUE (30) is excellent. Among basic works on military history are the last two volumes of the HISTOIRE UNIVERSELLE DES ARMEES (19), La Gorce's LA REPUBLIQUE ET SON ARMEE (21), the HISTOIRE POLITIQUE DE L'ARMEE (25), by Planchais and Nobécourt, and General Weygand's HISTOIRE DE L'ARMEE FRANCAISE (33). For the First World War there is also the massive official history, LES ARMEES FRANCAISES DANS LA GRANDE GUERRE (2).

THE HISTORY OF AVIATION

Here a number of general works may be cited, beginning with L'HOMME, L'AIR ET L'ESPACE, by Dollfus, Beaudois and Rougeron (14), which is valuable for its accounts of aerostation and early aviation. Other general histories by René Chambe (10) and Edmond Petit (23, 24) treat both military and civil aviation and cover events in France and in other countries as well. The history of the French aircraft industry to 1940 has been definitively treated in Emmanuel Chadeau's five-volume doctoral thesis "Etat, entreprise et developpement économique" (9). For French military aviation the fundamental work is the HISTOIRE DE L'AVIATION MILITAIRE FRANCAISE, produced by a team of scholars working under the direction of General Charles Christienne and General Pierre Lissarrague (12); this basic work has now appeared in English translation (13). Another useful work is that of General Hébrard (18), which covers the interwar period from the perspective of 1945. Also useful are the two volumes of André van Haute's PICTORIAL HISTORY OF THE FRENCH AIR FORCE (31). For French naval aviation the essential works are those of R. Vulliez (32) and Admiral Pierre Barjot (4).

PERIODICALS

A number of aeronautical journals contain valuable materials for the historian. First among them in time sequence is LA GUERRE AERIENNE, edited by Jacques Mortane from 1916 to 1919; it is a capital source for French aviation in the Great War. The prestigious French naval review entitled LA REVUE MARITIME has been a rich source for material on the Aéronavale from before the First World War up to the present. And for the years of its publication, 1921 to 1963, the journal LES AILES is of fundamental importance, both for the completeness of its reportage and for its editorial and analytical articles.

From January, 1921 until December, 1928, LA REVUE DE L'AERONAUTIQUE MILITAIRE was the organ for French military airmen, and an outlet for their reflections and their points of view. Besides various theoretical studies, the journal contained very precise information on the air arm, as well as its involvement in colonial wars in such places as Syria and Morocco. This journal was succeeded by the REVUE DES FORCES AERIENNES, the official organ of the new Air Ministry from 1929 to 1933. It was in turn replaced by the REVUE DE L'ARMEE DE L'AIR, which ran from 1933 to 1939; in the years 1935 and 1936 it was supplemented by a REVUE DU MINISTERE DE L'AIR.

Following World War II a new official journal appeared, the FORCES AERIENNES FRANCAISES, which began publication in 1946 and unfortunately ceased in 1971. This journal is extremely valuable to the historian for several reasons. It ranged in interest over the whole spectrum of military aviation, touching not only the past, but current themes and problems as well. It kept abreast of developments in other countries, and articles often used the comparative approach; at the same time it strove to keep its readers informed on developments in civil aviation, on technological matters, and on the literature appearing in the field of aeronautics. Finally, it was a remarkably open forum for ideas, thanks to its liberal editorial policy.

Another valuable postwar journal was the REVUE DE DEFENSE NATIONALE, which began to appear in 1945 and continues today. Its themes are the great problems of defense as they affect all three services, but developments in the air are well covered. Far more details on the Armée de l'Air are offered in the popular journal AIR ACTUALITES, which has been appearing since 1967. Official in nature and at the same time appealing to the broad mass of flying and ground personnel, it abounds in detailed information but lacks the scholarly and theoretical articles of FORCES AERIENNES FRANCAISES.

There are three other journals which are frankly historical in conception and often of great value to researchers in aviation history. The first of these is the REVUE HISTORIQUE DES ARMEES, which has undergone several changes since its first appearance in 1941 as the REVUE HISTORIQUE DE L'ARMEE. It is now the official historical journal of the three services, aimed at a broad readership. Then there is the REVUE D'HISTOIRE DE LA DEUXIEME GUERRE MONDIALE, which began publication in 1950 and remains the essential journal for the historian of the conflict of 1939-1945. And finally there is the journal ICARE, published by the Syndicat des Pilotes de Ligne Français, the French airline pilots' union, under the direction of Jean Lasserre with the help of Edmond Petit. Each number has a major theme, often presented through eyewitness accounts and striking illustrations; some themes explored have been the Battle of France, 1940, the air war from 1914 to 1918, and ballooning in the Franco-Prussian War.

The major newspapers reflect faithfully the various polemics that have arisen over aviation since its origins. This press is also a key to understanding public attitudes, which newspapers have both mirrored and influenced. Then too, newspaper columns often invoked the future of military aviation and the promises and menaces it seemed to offer. One could compare, for example, the great stir over aerochemical warfare in the 1930s with the discussion over the French nuclear striking force of the 1960s. Then there are the smaller provincial newspapers, which often reflect the life and the activities of air units stationed in the vicinity and give some indication of relations between aviation personnel and the local population. Here are rich possibilities for research, which have until now been largely unexplored.

BIBLIOGRAPHICAL LISTING

1. AIR UNIVERSITY LIBRARY INDEX TO MILITARY PERIODICALS. Washington: 1949 ff.

2. LES ARMEES FRANCAISES DANS LA GRANDE GUERRE. 105 vols. Paris: 1922-1937.

3. Azéma, Jean-Pierre. DE MUNICH A LA LIBERATION. Paris: 1979.

4. Barjot, Pierre. HISTOIRE DE LA GUERRE AERONAVALE. Paris: 1961.

5. Bernard, Philippe. LA FIN D'UN MONDE: 1914-1929. Paris: 1975.

6. BIBLIOGRAPHIE ANNUELLE DE L'HISTOIRE DE FRANCE. Paris: 1955 ff.

7. Bonnefous, Georges. L'HISTOIRE POLITIQUE DE LA IIIe REPUBLIQUE. 6 vols. Paris: 1956-1965.

8. Brockett, Paul. BIBLIOGRAPHY OF AERONAUTICS. Washington: 1909-1932.

9. Chadeau, Emmanuel. "Etat, entreprise et développement économique: l'industrie aéronautique en France (1900-1940)." Doctoral dissertation, Université de Paris X, 1986.

10. Chambe, René. HISTOIRE DE L'AVIATION. Paris: 1949.

11. Chastenet Jacques. HISTOIRE DE LA IIIe REPUBLIQUE. 7 vols. Paris: 1952-1963.

12. Christienne, Charles, Pierre Lissarrague, Alain Degardin, Patrick Facon, Patrice Buffotot and Marcellin Hodeir. HISTOIRE DE L'AVIATION MILITAIRE FRANCAISE. Paris: 1980.

13. _________. A HISTORY OF FRENCH MILITARY AVIATION. Washington: 1986.

14. Dollfus, Charles, Pierre Beaudois and Camille Rougeron. L'HOMME, L'AIR ET L'ESPACE. Paris: 1965.

15. Dubief, Henri. LE DECLIN DE LA TROISIEME REPUBLIQUE 1929-1938. Paris: 1976.

16. Duroselle, Jean-Baptiste. L'HISTOIRE DIPLOMATIQUE DE 1919 A NOS JOURS. Paris: 1957.

17. Goguel, François. LA POLITIQUE DES PARTIS SOUS LA IIIe REPUBLIQUE. 2 vols. Paris: 1946.

18. Hébrard, J. VINGT-CINQ ANNEES D'AVIATION MILITAIRE (1920-1945). 2 vols. Paris: 1946.

19. HISTOIRE UNIVERSELLE DES ARMEES. 4 vols. Paris: 1965-1966.

20. Köhler, Karl, BIBLIOGRAPHIE ZUR LUFTKRIEGSGESCHICHTE. Frankfort: 1966.

21. La Gorce, Paul-Marie de. LA REPUBLIQUE ET SON ARMEE. Paris: 1963.

22. Mayeur, Jean-Marie. LES DEBUTS DE LA TROISIEME REPUBLIQUE, 1871-1898. Paris: 1973.

23. Petit, Edmond. HISTOIRE MONDIALE DE L'AVIATION. Paris: 1967.

24. ________. NOUVELLE HISTOIRE DE L'AVIATION. Paris: 1966.

25. Planchais, Jean and Jacques Nobécourt. UNE HISTOIRE POLITIQUE DE L'ARMEE. 2 vols. Paris: 1967.

26. Rebérioux, Madeleine. LA REPUBLIQUE RADICALE? 1898-1914. Paris: 1975.

27. Renouvin, Pierre. LE XIXe SIECLE. 2 vols. Paris: 1954-1955.

28. ________. LES CRISES DU XXe SIECLE, I. DE 1914 A 1929. Paris: 1957.

29. Rioux, Jean-Pierre. LA FIN DE LA IVe REPUBLIQUE. 2 vols. Paris: 1980-1983.

30. Sauvy, Alfred. L'HISTOIRE ECONOMIQUE DE LA FRANCE ENTRE LES DEUX GUERRES (1919-1939). 4 vols. Paris: 1965-1975.

31. Van Haute, André. PICTORIAL HISTORY OF THE FRENCH AIR FORCE. 2 vols. London: 1974.

32. Vulliez, R. AERONAVALE (1915-1954). Paris: 1955.

33. Weygand, General Maxime. HISTOIRE DE L'ARMEE FRANCAISE. Paris: 1938.

THE BEGINNINGS, 1783-1914

For this long period which saw the birth and infancy of French military aviation the basic study is one produced before World War II by the Service des Etudes Historiques et Géographiques of the Air Ministry. This work, prepared under the direction of Colonel René Chambe, bears the title L'AVIATION MILITAIRE FRANCAISE AVANT LA GUERRE DE 1914. Unfortunately it cannot be listed with the usual bibliographical entries because it never appeared in print. It exists in typescript form, with a copy available for consultation at the S.H.A.A. The study traces in detail the first steps in the development of French military aviation and clearly shows the heightened interest in the heavier-than-air beginning in the last decade of the nineteenth century. The work also contains an important annex with the texts of relevant documents; these are all the more valuable since the originals were lost during World War II. Also available at the S.H.A.A. is the text of a 1985 paper by Patrick Facon entitled "L'Armée et l'aviation de 1891 à 1914," which shows that the Army's interest in aeronautics was real and continuous, and that in official circles a combat role for aircraft was clearly envisaged.

The role of the aerial pioneer Ferdinand Ferber can be traced in his work entitled L'AVIATION (54), and the fertile mind and prophetic thought of Clément Ader is mirrored in his L'AVIATION MILITAIRE (1). Ader has also been the subject of a study by the eminent British aviation historian Charles H. Gibbs-Smith (61), though not all authorities on early French aviation agree with Gibbs-Smith concerning Ader's claims to have achieved true flight. Useful for the early days of Army aviation are the articles George Bellenger published before World War II in the REVUE DE L'ARMEE DE L'AIR (13,14). For the earliest days of the Aéronavale and naval aviation generally, see the recent article by J.B. Lacroix in the REVUE MARITIME (71).

BIBLIOGRAPHICAL LISTING

1. Ader, Clément. L'AVIATION MILITAIRE. PARIS: 1909.

2. ________. LA PREMIERE ETAPE DE L'AVIATION MILITAIRE EN FRANCE. PARIS: 1907

3. "Aerial Battleships and Flying Torpedo Boats: German Dirigibles versus French Aeroplanes." SCIENTIFIC AMERICAN SUPPLEMENT No. 1959 (July 19, 1913), 35.

4. AEROSTIERS ET AVIATEURS. Paris: 1914.

5. Allard, Commandant. HISTOIRE DE LA NAVIGATION AERIENNE, BALLONS, AEROPLANES. Besançon: 1909.

6. Armengaud, J. LE PROBLEME DE L'AVIATION. SA SOLUTION PAR L'AEROPLANE. Paris: 1908.

7. L'AVIATION. SES DEBUTS, SON DEVELOPPEMENT. Paris: 1908.

8. L'AVIATION TRIOMPHANTE. Preface by M. d'Estournelles de Constant. Paris: 1909.

9. AVIONNERIE MILITAIRE, POINTAGE AERIEN: INSTRUMENTS DE MESURES POUR AVIONS TORPILLEURS. Paris: 1912.

10. Bague, Lieutenant E. MES PREMIERES IMPRESSIONS D'AVIATEURS. Paris: 1911.

11. Baudry de Saunier (L). ELEMENTS DE LOCOMOTION AERIENNE. Paris: 1909

12. Beaubois, Henry. "Un cinquantenaire: Le premier concours d'avions militaires (Reims: octobre-novembre 1911)." F.A.F. 176 (Dec. 1961), 33-42.

13. Bellenger, Geerges. "Notes et souvenirs sur d'aviation militaire avant 1914." R.A.A. (Feb. 1938), 143-180.

14. ________. Notes et souvenirs sur la création de l'aviation militaire française." R.A.A. 79 (Feb. 1936), 167-197; 82 (May 1936), 512-534; 83 (June 1936), 688-698; 84 (July 1936), 795-802; 85 (Aug.

1936), 917-924; 86 (Sept. 1936), 1025-1038; 87 (Oct. 1936), 1162-1174; 88 (Nov. 1936), 1241-1262; 89 (Dec. 1936), 1374-1390; 93 (April 1937), 390-422; 94 (May 1937, 533-562).

15. Bellenger, Jean. LA GUERRE AERIENNE ET LE DROIT INTERNATIONAL. Paris: 1912.

16. Benichou, M. "Mise en place d'une doctrine d'emploi pour l'arme aérienne par l'Armée française. 1911-1912." COLLOQUE INTERNATIONAL. ADAPTATION DE L'ARME AERIENNE AUX CONFLITS CONTEMPORAINS ET PROCESSUS D'INDEPENDANCE DES ARMEES DE L'AIR DES ORIGINES A LA FIN DE LA SECONDE GUERRE MONDIALE. Paris: 1985.

17. Berget, Alphonse. BALLONS, DIRIGEABLES ET AEROPLANES. Paris: 1908.

18. Berrubé, Emile. FLOTTES AERIENNES. EN FRANCE ET EN ALLEMAGNE. AEROPLANES ET BALLONS DE GUERRE. LOIS DE L'AERO-DYNAMIQUE. LE DIRIGEABLE RATIONNEL. TORPILLES AERIENNES. Paris: 1910.

19. Bertrand, Capitaine Charles. ETAT ACTUEL DE L'AERONAUTIQUE MILITAIRE ET NAVALE EN FRANCE ET A L'ETRANGER. Paris: 1914.

20. Besançon, Georges. BALLONS ET AEROPLANES. Paris: 1910.

21. Besseyre des Horts, Commandant. ESSAI SUR L'UTILISATION DU DIRIGEABLE ET DE L'AEROPLANE EN CAMPAGNE. Paris: 1910.

22. Bethuys, G. LES AEROSTIERS MILITAIRES. Paris: 1889.

23. Bienvenue, Capitaine. LA BALLON A CASABLANCA (1907-1908). Paris: 1908.

24. Blanc, Edmond. "Le colonel Charles Renard." L.A. 1118 (June 28, 1947), 4-5.

25. Bois, Contrôleur-Général. "Les Ballons dirigeables à Chalais Meudon de 1906 à 1911." F.A.F. 154 (Dec. 1959), 1029-1058.

26. Boisricheux, Lieutenant de. LES AEROPLANES DANS LA

GUERRE D'AUJOURD'HUI. Paris: 1912.

27. Bol'shev, M.N. MORSKOE VOZDUKHOPLAVANIE VO FRANTSII. Saint Petersburg: 1904.

28. Bonnefoy, Capitaine. "Conceptions successives de l'avion de guerre: un document de 1913." R.F.A. 19 (Feb. 1931), 168-187.

29. Booms, Wilhelm. "Die Tätigkeit des Flugmaschinenparks bei den grossen französischen Manövern 1912." OESTERREICHISCHE FLUG-ZEITSCHRIFT 6 (Jan. 10, 1913), 11-15.

30. Bornecque, J. L'AEROSTATION MILITAIRE EN FRANCE ET A L'ETRANGER. Paris: 1899.

31. Boullaire, Capitaine. "Aéroplanes et cavalerie." R.M.G. (July 1911), 24.

32. Boye, Pierre. LES PREMIERES EXPERIENCES AEROSTATIQUES FAITES EN LORRAINE (1783-1788). Paris: 1909.

33. Branche, Joseph. LES BALLONS D'OBSERVATION. Paris: 1977.

34. Cahisa, R. L'AVIATION D'ADER ET DES TEMPS HISTORIQUES. Preface by R. Morane. Paris: 1950.

35. Cantelou, Maurice de. ETUDE SUR L'AVIATION. Paris: 1909.

36. Castey, L. L'HOMME QUI DONNA DES AILES AU MONDE: CLEMENT ADER. Paris: 1947.

37. Chaffal, Jean du. "Avec l'aviation au Maroc au temps héroique de la pacification." F.A.F. 154 (Dec. 1959), 1087-1102.

38. Challeat, J. ARMEES MODERNES ET FLOTTES AERIENNES. Paris: 1911.

39. Charet, Capitaine. "L'aviation d'artillerie." REVUE D'ARTILLERIE (Mar. 1912), 27.

40. "Clément Ader, créateur et prophète." L.A. 1134 (Oct. 18, 1947), 1-4.

41. Conneau, Jean-Louis-Camille. MY THREE BIG FLIGHTS. London: 1912.

42. CONCOURS D'APPAREILS MILITAIRES D'AVIATION. Paris: 1913.

43. Courty, Jacques. L'AEROSTATION ET SES APPLICATIONS MILITAIRES. Paris: 1900.

44. Debuchy, Victor. LES BALLONS DU SIEGE DE PARIS. Paris: 1973.

45. Delon, C. PROMENADES DANS LES NUAGES: Paris: 1881.

46. Dollfus, Charles. EN BALLON. Paris: 1962.

47. Dollfus, Charles, Paul Maincent and André Guerin. "La Guerre de 1870-1871." ICARE 56 (1971)("Les ballons, le poste aérienne"); 77 (1976)("Les aérostiers, les dirigeables").

48. Duhem, Jules. HISTOIRE DE L'ARME AERIENNE AVANT LE MOTEUR. Paris: 1964.

49. Espitallier, G. LES BALLONS ET LEUR EMPLOI A LA GUERRE. Paris: 1889.

50. Estienne, Colonel J.E. "La valeur militaire de l'aviation." REVUE D'ARTILLERIE (Jan. 1913), 14.

51. Etévé, Albert. LA VICTOIRE DES COCARDES. L'AVIATION FRANCAISE AVANT ET PENDANT LA PREMIERE GUERRE MONDIALE. Preface by A. Caquot. Paris: 1970.

52. "Les études d'armament aérien en 1910-1911." R.A.A. 105 (April 1938), 398-424.

53. Fêrber, Andrée. LES DEBUTS VERITABLES DE L'AVIATION FRANCAISE. Paris: 1970.

54. Ferber, Ferdinand. L'AVIATION; SES DEBUTS - SON DEVELOPPEMENT. Paris: 1910.

55. _________. LES PROGRES DE L'AVIATION DEPUIS 1891 PAR LE VOL PLANE. Paris: 1905.

56. Fontvielle, Wilfrid de and Georges Besançon. NOTRE

FLOTTE AERIENNE. Paris: 1908.

57. "Französische Militärflugzeuge." ZEITSCHRIFT FUER FLUGTECHNIK UND MOTORLUFTSCHIFFAHRT 3 (Oct. 12, 1912), 253-255.

58. "French Aeroplane Designs." AERO 6 (Feb. 1912) 48-50.

59. "The French Military Aeroplane Trials." AERO 5 (Nov. 1911), 211-217.

60. Frey, Général H. L'AVIATION AUX ARMEES ET AUX COLONIES ET AUTRES QUESTIONS MILITAIRES ACTUELLES. Nancy: 1911.

61. Gibbs-Smith, Charles H. CLEMENT ADER: HIS FLIGHT-CLAIMS AND HIS PLACE IN HISTORY. London: 1968.

62. Hamon, James. MANUEL DE L'AVIATEUR. NOTIONS GENERALES SUR L'AVIATION A L'USAGE DES AVIATEURS ET DES CANDIDATS AUX TROUPES DE L'AVIATION MILITAIRE. Paris: 1913.

63. Heide, F. "Die Organisation and Stärke des französischen Militärluftfahrtwesens." DEUTSCHE LUFTFAHRER-ZEITSCHRIFT 17 (June 25, 1913), 318-320.

64. Hirschauer, Commandant. INSTRUCTION PRATIQUE SUR LES AEROSTATS MILITAIRES. n.p.: 1903.

65. HISTOIRE DES BALLONS ET DES AERONAUTS CELEBRES. Paris: 1980.

66. "Histoire des forces aériennes françaises." ICARE 91 (1979-1980), 3-69.

67. Joanneton, Henry. NOTE SUR L'UTILISATION DES AEROPLANES POUR LE LANCEMENT DES PROJECTILES. Troyes: 1912.

68. Josse, R. "Ferdinand Ferber, 1862-1909: Pionnier de l'aviation." R.H.A. (No. 2, 1974), 112-132.

69. Jourdain, Pierre-Roger. "Aviation in France in 1908." SMITHSONIAN INSTITUTION ANNUAL REPORT 1908, 145-159.

70. Joux, Etienne. UN DIRIGEABLE MILITAIRE: L'ADJUDANT VINCENNOT (1911-1916). Paris: 1931.

71. Lacroix, J.B. "Naissance de l'aéronautique navale française." REVUE MARITIME 341 (Nov. 1978), 2505-2518; 342 (Dec. 1978), 2623-2635.

72. Lafargue, Lieutenant de. L'AVIATION PAR UN CAVALIER. Paris: 1911.

73. Lafon, Charles. L'AERONAUTIQUE NAVALE MILITAIRE MODERNE. Paris: 1914.

74. ________. CONFERENCES SUR L'AVIATION MILITAIRE ET NAVALE. Paris: 1912.

75. Lainé, Berge. "L'Aéronautique militaire française au Maroc (1911-1939)." R.H.A. No. 4 (1978), 107-119.

76. Lancy, H. L'AVIATION DE SES ORIGINES A 1914. Lyon: 1939.

77. Laux, James M. "The Rise and Fall of Armand Deperdussin." FRENCH HISTORICAL STUDIES 8 (Spring 1973), 95-104.

78. Maincent, Paul. "L'aérostation d'observation à Saint-Denis en 1870." F.A.F. 143 (Dec. 1958), 909-920.

79. Mallet, François. LA CONQUETE DE L'AIR ET LA PAIX UNIVERSELLE. Paris: 1910.

80. MANUEL DE L'AERONAUTIQUE MILITAIRE. Paris: 1912.

81. Marchise, L. L'EPOPEE AERIENNE. Paris: 1910.

82. Marie, Colonel Félix. LES ORIGINES DE L'AERONAUTIQUE MILITAIRE. (NOVEMBRE 1909-NOVEMBRE 1910). Paris: 1924.

83. Mellor, Clive. THE AIRMAN. EXPERIENCES WHILE OBTAINING A BREVET IN FRANCE. London: 1913.

84. "Die Militäraviatik in Frankreich." WIENER LUFTSCHIFF-FUEHRER ZEITUNG 10 (Sept. 15, 1911), 331-333.

85. Miller, Warren H. "The Manufacture of French Airplanes." ENGINEERING MAGAZINE 39 (Aug. 1910), 649-658.

86. Mortane, Jacques. "The Evolution of Military Aviation in France." Tr. C. de F. Chandler. FLYING (N.Y.) I (Dec. 1912), 26-29; (Jan. 1913), 13-15, 30-32.

87. NOTRE AVENIR EST DANS L'AIR. Sceaux: 1912.

88. Omezon, Léopold d'. CE QU'IL FAUT SAVOIR DE L'AVIATION. SON ROLE AUX ARMEES. Paris: 1910.

89. "The Organization of Military Aeronautics in France." AERONAUTICS (London) 4 (June, 1911), 87-88.

90. Painlevé, Paul and Emile Borel. L'AVIATION. Paris: 1910.

91. Payen, Jacques. LES PREMIERS AEROSTATS (1783-1804). Paris: 1967.

92. Perreau, Joseph. L'AEROFLOTTE, AILE DE LA VICTOIRE. PLAN D'ORGANISATION TACTIQUE ET D'APPLICATION STRATEGIQUE DE L'AERONAUTIQUE MILITAIRE. Paris: 1913.

93. Petit, Edmond. LA VIE QUOTIDIENNE DANS L'AVIATION EN FRANCE AU DEBUT DU XX SIECLE (1900-1935). Paris: 1977.

94. Philit, Edmond. LA GUERRE AERIENNE. Montpellier: 1910.

95. Pichot-Duclos, Capitaine. RECONNAISSANCES EN AEROPLANE. THEORIE; CAS CONCRETS, PROPOSITIONS. Paris: 1912.

96. Poydenot, H. QUAND LE CIEL ETAIT FRANCAIS. 1783-1913. Paris: 1956.

97. "Les questions d'organisation, après les manoeuvres de 1910." R.A.A. 107 (June 1936), 625-666.

98. Remy, Lieutenant. COMMENT ON FORME UN AVIATEUR. Preface by H. Farman. Paris: 1912.

99. Renard, Commandant P. L'AVIATION. Paris: 1909.

100. Reymond, Emile. INTERPELLATION DE M. LE DOCTEUR EMILE REYMOND, SENATEUR...SUR L'AEROSTATION MILITAIRE. Paris: 1910.

101. Richepin, Jean. L'AILE. ROMAN DES TEMPS NOUVEAUX. Paris: 1911.

102. LA ROUTE DE L'AIR. AERONAUTIQUE, AVIATION, HISTOIRE, THEORIE, PRATIQUE. Paris: 1909.

103. Sensever, G. and L. Ballif. LE COMBAT AERIEN. Paris: n.d.

104. Tatin, V. THEORIE ET PRATIQUE DE L'AVIATION. Paris. 1910.

105. Thomas, M.R. "Les premiers raids aériens en Afrique: 1911-1933. F.A.F. 165 (Decembre 1960), 739-768.

106. Timonier, Pol. COMMENT NOUS TORPILLERONS BERLIN AVEC NOTRE ESCADRILLE D'AEROPLANES DES L'OUVERTURE DES HOSTILITES. Paris: 1913.

107. Tissandier, Gaston. SOUVENIRS ET RECITS D'UN AEROSTIER MILITAIRE. (1870-1871). Paris: 1891.

108. Turgan, Louis. HISTOIRE DE L'AVIATION, AVIONS ET AVIATEURS D'HIER, D'AUJOURD'HUI, DE DEMAIN (ETUDE DOCUMENTAIRE). Paris: 1909.

109. Vallet, Albert. L'AVIATION MILITAIRE EN 1912. SON ROLE DANS LA GUERRE MODERNE. Paris: 1913

110. Woodhouse, Henry. "France's National Movement to Gain Aerial Supremacy." AERO CLUB OF AMERICA BULLETIN 1 (May 1912), 13-16.

There is no dearth of literature on the aerial aspects of the Great War, but much of that literature leaves something to be desired; it is often anecdotal or hagiographical in nature. In one category of writing the period 1914-1918 is far richer than the preceding period, and that is in scholarly studies coming from universities. Among the most promising of academic historians was Philippe Bernard, whose unfortunate death in 1969 prevented the completion of his thesis, though he has left a provocative article on air strategy (27). In a sense his work has been continued by Captain Jean-Marc Marrill, who defended his thesis on the birth of French military aviation at the University of Paris I in December, 1985 (156). This work of some five hundred pages lays greater stress on the tactical aspect than on political or strategic facets of the subject; still in all it is a work of great usefulness and fills a gap in our knowledge.

Among other works of special value is the massive AERONAUTIQUE PENDANT LA GUERRE MONDIALE edited by Maurice de Brunoff (43) in 1919. Despite its age it offers a useful panorama of the French Air Service and its activities. General Pierre Lissarrague also provides much information in his MEMENTO (141). The testimony and the analyses presented by Simone Pesquiès-Courbier in the pages of the review ICARE (178) shed considerable new light on the subject. The problems of doctrine for the new arm are treated in DANS LES COULISSES, by Georges Huisman (110), which retains its usefulness even though it appeared not long after the Armistice of 1918. Matters of doctrine are also treated by General Voisin whose work (238) is an affirmation of faith in the subordination of the air arm to the ground forces, and whose objectivity suffers accordingly. Also useful in this connection in Patrick Facon's study of the air division (84).

Turning to various types of aviation the monumental work of René Martel on bombardment aviation (157) remains the basic work on the subject, though it can be completed by the study of Simone Pisquès-Corbier (200) on the bombardment of the Briey Basin. As for fighter aviation, it has recently been studied for the years 1917-1918 by Jérôme Blanc (30) in a thesis completed at the University of Paris I in 1982. Among histories of flying units can be cited the recent study of Groupe de Bombardement 1 by Simone Pesquiès-Courbier (199) and the history of Groupe de Chasse

I/2, the "Storks," by Patrice Buffotot and Yves Brèche (45). There is a considerable amount of literature on the Lafayette Escadrille, but the most recent and comprehensive study is THE VIVID AIR by Philip Flammer (88).

The story of aircraft production during the war is well covered in the thesis of Emmanuel Chadeau, cited earlier in the general guide to published sources, but for the period of the Great War it can be supplemented by more specialized studies, such as that by Jean Liron on LES AVIONS FARMAN (140). The procurement and training of airmen has been the subject of a number of works including the thesis of Odile Benoist (25). The thesis of Marcellin Hodeir (109) develops this subject more completely both in its statistics and in its probing of the airman's mental makeup. Jean-Pierre Dournel has examined the image of the flyer as it appeared in one of the more popular reviews of the era (76) and Daniel Porret has prepared capsule biographies of the 187 French aces (208) which open the door to all sorts of comparative and analytical studies.

BIBLIOGRAPHICAL LISTING

1. Ader, Clément. POUR LA SECURITE ET L'HONNEUR DE LA FRANCE. DES VERITES SUR L'UTILISATION DE L'AVIATION MILITAIRE AVANT ET PENDANT LA GUERRE. UNE FAUTE DESASTREUSE, SUPREME LECON POUR L'AVENIR! Toulouse, Douladoure, 1919.

2. "L'aérostation dans la guerre de 1914-1918." ICARE 83 (1978), 114-161.

3. Ajalbert, Jean. L'AVIATION AU-DESSUS DE TOUT. Paris: 1916.

4. _________. LA PASSION DE ROLAND GARROS. Paris: 1934.

5. Allard, Paul. IMAGES SECRETES DE LA GUERRE: 200 PHOTOGRAPHIES CENSUREES EN FRANCE. Paris: 1933.

6. D'Ami, Rinaldo. "L'aviazione francese alla difesa di Venezia nella prima guerra mondiale." IL NUOVO CORRIERE FILATELICO 6 (1967), 240-259.

7. André, C.H.A. ABOVE THE CLOUDS. Trans. Mrs. Philip Duncan Wilson. Garden City, NY: 1919.

8. _________. AU DESSUS DES BATAILLES. CARNET DE GUERRE D'UN AVIATEUR. Paris: 1917.

9. André, Eric. "Le Général Rocques et l'inspection permanente de l'aéronautique." Unpublished mémoire de maîtrise, Université Paul Valéry, Montpellier, 1985.

10. Angot, E. and R.de Lavergne. UNE FIGURE LEGENDAIRE DE L'AVIATION FRANCAISE DE 1914 A 1940. LE GENERAL VUILLEMIN. LE COMBATTANT. LE PREMIER DU SAHARA. LE CHEF. Paris & Geneva: 1965.

11. L'ARMEE DE L'AIR. SA PREDOMINANCE ET SA TACTIQUE. Paris: 1915.

12. Armengaud, Général. LE RENSEIGNEMENT AERIEN, SAUVEGARDE DES ARMEES. Preface by Général Castelnau. Paris: 1934.

13. Arnoux, Jacques d'. PAROLES D'UN REVENANT. Preface by Henry Bordeaux. Paris: 1925.

14. LES AS PEINTS PAR EUX-MEMES. ETUDE SUR LES HEROS DESPARUS, SUIVIE D'ANECDOTES RECUEILLIES PAR M. JACQUES MORTANE. Paris: 1917.

15. Aubigny, d'. RAPPORT SUR LES TRAVAUX DE LA COMMISSION DE L'ARMEE PENDANT LA GUERRE DE 1914-1918: AERONAUTIQUE. Paris: 1919.

16. Azorin. PARIS BOMBARDEO. MAYO Y JUNIO 1918. Madrid: 1921.

17. Badel, Emile. LES BOMBARDEMENTS DE NANCY, VILLE OUVERTE. Nancy: 1919.

18. Barzini, Luigi. LA GUERRE MODERNE SUR TERRE, DANS LES AIRS ET SOUS LES EAUX. Paris: 1917.

19. Battesti, François. LES CIGOGNES DE BROCARD AU COMBAT. Paris: 1975.

20. Baume, Renaud de la. "L'aviation de bombardement 1914-1916." A.A. 321 (May, 1979), 42-48.

21. Bédier, Joseph. L'EFFORT FRANCAIS. QUELQUES ASPECTS DE LA GUERRE. Paris: 1919.

22. Bellenger, Lieutenant-Colonel Georges. "La bataille de la Marne vue d'en haut." ICARE 32 (1965), 16-39.

23. Belleroche, Pierre. HISTOIRE DU COMBAT AERIEN. Marseille: 1941.

24. Bellet, D. and W. Darville. LA GUERRE MODERNE ET SES NOUVEAUX PROCEDES. Paris: 1916.

25. Benoist, Odile. "Le recrutement des aviateurs français pendant la guerre de 1914-1918." Unpublished mémoire de maîtrise, Université de Paris I, 1975.

26. Beraud-Villars, Jean. NOTES OF A LOST PILOT. Trans. and ed. by Stanley J. Pincetti. Hamden, Conn.: 1975. See entry 147.

27. Bernard, Philippe. "A propos de la stratégie aérienne pendant la Première Guerre Mondiale: mythes et réalités." R.H.M.C. 16 (July-Sept., 1969), 350-375.

28. Biddle, Charles John. THE WAY OF THE EAGLE. New York: 1919.

29. Blaise, Lieutenant-Colonel M. L'EXPERIENCE DE LA GUERRE. L'EXPLORATION AERIENNE A LA III ARMEE EN AOUT-SEPTEMBRE 1914. Paris: 1929.

30. Blanc, Jérôme. "L'aviation de chasse française en 1917-1918." Unpublished mémoire de maîtrise, Université de Paris I, 1982.

31. "Le bombardement de Paris." AVENIR ECONOMIQUE ET FINANCIER, Feb. 28, 1919, 291-313.

32. Bommart, J. and J. Puistienne. ESCADRILLE 155. D'APRES LES NOTES DE GUERRE DU PILOTE JEAN PUISTIENNE. Preface by G. Sardier. Paris: 1935.

33. Bonnefon, Paul. LE PREMIER "AS", PEGOUD. Preface by Colonel Girod. Paris: 1918.

34. Bordeaux, Henry. LE CHEVALIER DE L'AIR: GUYNEMER. Paris: 1938.

35. _________. GEORGES GUYNEMER, KNIGHT OF THE AIR. Trans. by Louise Morgan Sill. New Haven: 1918.

36. _________. LA VIE DU GUYNEMER. Paris: 1920.

37. _________. VIE HEROIQUE DE GUYNEMER: LE CHEVALIER DE L'AIR. Paris: 1918.

38. Bordecque, H. and René de Valfosi. LES AILES DANS LA BATAILLE. Paris: 1920.

39. Boulenger, Jacques. EN ESCADRILLE. Paris: 1930.

40. Bourcier, Emmanuel. LE BOMBARDIER CAMUS. Paris: 1919.

41. Bourget, Général P.A. LE GENERAL ESTIENNE: PENSEUR, INGENIEUR ET SOLDAT. Preface by Général Weygand. Paris: 1956.

42. Bowen, Robert Sidney. THEY FLEW TO GLORY; THE STORY OF THE LAFAYETTE FLYING CORPS. New York: 1965.

43. Brunoff, Maurice de (Ed.). L'AERONAUTIQUE PENDANT LA GUERRE MONDIALE. Paris: 1919.

44. Buffotot, Patrice. "Les insignes de l'Armée de l'Air: un élément de la symbolique militâire." In Service Historique de l'Armée de l'Air, RECUEIL D'ARTICLES ET D'ETUDES (1974-1975). Vincennes: 1977, 1-18.

45. Buffotot, Patrice and Yves Brèche. HISTORIQUE DU GROUPE DE CHASSE I/2, LES CIGOGNES. Vincennes: 1982.

46. Cagny, J. de. "Aérostiers de 1914-1918." R.H.A. (No. 2, 1967), 69-90.

47. Cahisa, Raymond. DEUX GUERRES AERIENNES. QUE SERA CELLE DE DOMAIN? Paris: 1947.

48. Carlier, A.H. LA PHOTOGRAPHIE AERIENNE PENDANT LA GUERRE. Paris: 1921.

49. Castex, Marcel. ETUDE PHYSIO-PATHOLOGIQUE DE L'AVIATION DE GUERRE. Paris: 1921.

50. Chack, P. LA ROUTE DES INDES SAUVEE PAR LA FRANCE. Paris: 1941.

51. Chambe, René. AU TEMPS DES CARABINES. Paris: 1955.

52. ________. DANS L'ENFER DU CIEL. Paris: 1947.

53. ________. GUYNEMER. Paris: 1949.

54. ________. SOUS LE CASQUE DE CUIR. Preface by M. Renard. Paris: 1947.

55. Chaput, Jean. QUELQUES LETTRES DE JEAN CHAPUT, LIEUTENANT PILOTE COMMANDANT L'ESCADRILLE SPA 57 (1914-1918). Preface by Général Perruchon. Paris: n.d.

56. Chatelain, Capitaine. CAUSERIE PRATIQUE AUX ELEVES PILOTES. Chartres: 1918.

57. Chavagnes, René de. L'AVIATION DE CHASSE. LE GROUPE DES CIGOGNES. DE GUYNEMER A FONCK. Preface by Paul Painlevé. Paris: 1920.

58. Chevennes, F. LA GLOIRE A DES AILES. Preface by Louis Breguet. Paris: 1932.

59. Christienne, Général C. "Le dēbat de doctrine sur l'emploi de l'aviation militaire française (1916-1946)." R.D.N. June, 1979, 109-120.

60. Claudin, Jean. NOS AVIONS. DUELS EMOUVANTS DANS LES AIRS. LE TRIOMPHE DE NOS OISEAUX. Paris: 1915.

61. Cocteau, Jean. DANS LE CIEL DE LA PATRIE. Paris: 1918.

62. Conze, Général d'armée P. "Essai sur l'artillerie antiaérienne." R.H.A. (no. 1-2, 1975), 132-148.

63. Corneloup, Joannis. SOUVENIRS DE'UN AEROSTIER DE LA GRANDE GUERRE. Paris: 1975.

64. Cossart, Jean. LES HALLUCINATIONS DU LIEUTENANT DARNOUX. Paris: 1936.

65. Croidys, Pierre. GUYNEMER. Paris: 1939.

66. Crouvezier, Gustave. L'AVIATION PENDANT LA GUERRE. Preface by Barrès. Paris: 1915.

67. ________. LA GUERRE AERIENNE. LE ROLE DE LA 5° ARME. Paris: 1916.

68. Daurit. LA GUERRE EN BALLONS. GUERRE DE DEMAIN, GRAND RECIT PATRIOTIQUE ET MILITAIRE. Paris: 1921.

69. Delacommune, Charles. L'ESCADRILLE DES EPERVIERS. IMPRESSIONS VECUES DE GUERRE AERIENNE. Preface by M. Barrès. Paris: 1918.

70. Delanney, Lieutenant-Colonel. "Le sort de nos observateurs en avion pendant la guerre de 1914-1918." R.F.A. (June-Dec., 1928), 416-435.

71. Demaizière, L. de. UN GRAND PILOTE: ROMANET. Paris: 1980.

72. Descaves, Lucien. DANS PARIS BOMBARDE (1871-1914-1918). Paris: 1918.

73. Desgrandschamps, Georges. IMPRESSIONS D'UN BOMBARDIER (1915). Paris: 1931.

74. Desvosges, Jacques. UN REVE D'AVANT-GUERRE. JOURNAL D'UNE FRANC-AVIATEUR. Paris: 1917.

75. Dollfuss, Charles. "Les plus légers que l'air dans la bataille." ICARE 32 (1965), 45-51.

76. Dournel, J.P. "L'image de l'aviateur français en 1914-1918. Une étude du milieu des aviateurs d'après la revue 'La Guerre Aérienne'." R.H.A. (No. 4, 1975), 59-84; (No. 1, 1976), 95-123.

77. Duval, Général. "Pourquoi fut créée la division aérienne." R.A.A. 76 (Nov., 1935), 1270-1274.

78. Duval, Jacques. L'ARMEE DE L'AIR. L'AVION, LES AVIATEURS; AVIATION D'OBSERVATION, DE BOMBARDEMENT, DE CHASSE. Preface by Général Duval. Paris: 1918.

79. Duvau, André. BR 29. SOUVENIRS D'ESCADRILLE. Ed. by P. Debofle. Vincennes: 1976.

80. "Emploi de l'aérostation pendant l'offensive de septembre-octobre 1918 sur le front de la 4° armée." R.A.M. 11 (Sept.-Oct., 1922), 101-107.

81. Escoffier, Jean. GUYNEMER. Monaco: 1958.

82. Etévé, Albert. LA VICTOIRE DES COCARDES. L'AVIATION FRANCAISE AVANT ET PENDANT LA PREMIERE GUERRE MONDIALE. Preface by A. Caquot. Paris: 1970.

83. Evett Donald. "The World's First Ace [Roland Garros]." A.H. 6 (July, 1964), 87-88.

84. Facon, Patrick. "Un exemple de l'adaptation de l'arme aérienne aux conflits contemporains: la division aérienne de 1918." COLLOQUE INTERNATIONAL SUR L'ADAPTATION DE L'ARME AERIENNE AUX CONFLITS CONTEMPORAINS ET PROCESSUS D'INDEPENDANCE DES ARMEES DE L'AIR. Paris: 1985, 223-238.

85. Farré, Henry. SKY FIGHTERS OF FRANCE; AERIAL WARFARE 1914-1918. Trans. by Catharine Rush. Boston: 1919.

86. Flammer, Phillip. "Lufbéry: Ace of the Lafayette Escadrille." A.H. 3 (Jan., 1961), 13-22.

87. ________. "The Myth of the Lafayette Escadrille." A.H. 22 (Spring, 1979), 23-28.

88. ________. THE VIVID AIR. THE LAFAYETTE ESCADRILLE. Athens, Ga.: 1981.

89. Fleury, Colonel Jean. "Dix ans après." R.D.N. June, 1979, 71-80.

90. Florian, Parmentier. LE ROMAN DE NUNGESSER. Paris: 1946.

91. Fonck, René. ACE OF ACES. Preface by Marshal Foch. Ed. by Stanley M. Ulanoff. Trans. by Martin M. Sabis and Stanley M. Ulanoff. Garden City, N.Y.: 1967.

92. ________. MES COMBATS. Paris: 1920.

93. Franc-Nohain, and Paul Delay. LA GUERRE AERIENNE. Paris: 1917.

94. "French Airships." AEROPLANE Aug. 20, 1919, 750-752.

95. Fridenson, Patrick. "Bilan des forces aériennes françaises." F.A.F. 252 (Nov., 1969), 473-489.

96. Garsonnin, Capitaine. "Commentaires et souvenirs d'un pilote de l'aviation francaise des Dardannelles." R.A.A. 90 (Jan., 1937), 17-36; 91 (Feb., 1937), 153-184; 92 (Mar., 1937), 313-348.

97. Garros, Roland. MEMOIRES. Paris: 1966.

98. Genet, E.C.C. AN AMERICAN FOR LAFAYETTE: THE DIARIES OF E.C.C. GENET, LAFAYETTE ESCADRILLE. Ed. by Walt Brown, Jr. Charlottesville, Va.: 1985.

99. Goutard, A. "Nos ailes en 1914." HISTORIA 306 (May, 1972), 88-104.

100. Gouvieux, Marc. HAUT LES AILES! CARNET DE ROUTE D'UN OFFICIER AVIATEUR PENDANT LA GUERRE DE 19... Paris: 1914.

101. _________. NOTES D'UN OFFICIER OBSERVATEUR EN AVION. Paris: 1916.

102. Grard, Lieutenant-Colonel. L'ACIER, AVIATION, AUTOMOBILISME, CONSTRUCTION MECANIQUE, SANCTIONS DE LA GUERRE. Paris: 1919.

103. Guillemeney, Colonel. "Le bombardement aérien des installations industrielles. La blocus du bassin de Briey." R.F.A. 15 (Oct., 1930), 1151-1190.

104. Hall, J.N. and C.B. Nordhoff. THE LAFAYETTE FLYING CORPS. New York: 1920.

105. Harcourt, Général J. d'. "Une escadrille de reconnaissance en 1914-1915." R.H.A. (No. 3, 1964), 119-124.

106. Hennessy, Juliette. "The Lafayette Escadrille - Past and Present." A.H. 3 (July, 1957), 150-161.

107. Hie, Serge. LES PREMIERS CHEVALIERS DU CIEL. 52 AS DE L'AVIATION DE CHASSE FRANCAISE DE 1914-1918. Paris: 1973.

108. HISTORIQUE DU 21° REGIMENT D'AVIATION (1914-1918). Paris: 1924.

109. Hodeir, Marcellin. "L'aviation militaire française de la première guerre mondiale: 1917-1918. Etude du milieu social, approche des mentalités." Unpublished mémoire de maîtrise, Université de Paris IV, 1977.

110. Huisman, Georges. DANS LES COULISSES DE L'AVIATION: 1914-1918. POURQUOI N'AVONS NOUS PAS TOUJOURS GARDE LA MAITRISE DES AIRS? Paris: 1921.

111. Ide, John J. "The Development of French Naval Aviation." AVIATION 7 (Nov. 15, 1919), 340-345.

112. Jannekeyn, Commandant. "Un épisode de la 'ruée sur

Paris'. La contre attaque Mangin du 11 juin 1918." R.F.A. (Aug., 1929), 41-64.

113. Jeanjean, Marcel. DES RONDS DANS L'AIR. Introduction by Général R. Barthélémy. Aurillac: 1967.

114. Joscat, René. SIRENE D'ESCADRILLES. Paris: 1932.

115. Jullian, Marcel. LE CHEVALIER DU CIEL. CHARLES NUNGESSER. Preface by R. Nungesser. Paris: 1953.

116. ________. GENS DE L'AIR. Saint-Amand: 1959.

117. ________. "L'ouverture de la Chasse. Frantz et Quesnault contre Aviatik." ICARE 32 (1965), 53-56.

118. ________. LA GRANDE BATAILLE DANS LES AIRS, 1914-1918. Paris: 1967.

119. Kessel, Joseph. LE REPOS DE L'EQUIPAGE. Paris: 1935.

120. ________. L'EQUIPAGE. Paris: 1942.

121. ________. LES TEMPS SAUVAGES. Paris: 1976.

122. Labric, Roger. JEAN DAGNAUX, L'AS A LA JAMBE DE BOIS. Paris: 1947.

123. Lacroix, F. EN PLEIN CIEL: IMPRESSIONS D'AVIATEUR, SENSATIONS DE VOL, LA GUERRE EN AVION. Paris: 1918.

124. Lacroix, Pierre. LE DOMAINE AERIEN ET LA GUERRE. ESSAI SUR LES HOSTILITES ET LA NEUTRALITE AERIENNES DANS LE CONFLIT DE 1914-1919. Toulouse: 1919.

125. La Fargue, Commandant de. "L'aviation militaire en Tunisie. Historique. Role (1916-1920)." R.A.M. 3 (1921), 49-58.

126. Laffray, Jean. PILOTE DE CHASSE AUX CIGOGNES. Paris: 1968.

127. Lafon, Charles. LA FRANCE AILEE EN GUERRE. Paris: 1920.

128. ________. LES ARMEES AERIENNES MODERNES. FRANCE ET ETRANGER. Paris: 1916.

129. Lafont, Bernard. AU CIEL DE VERDUN. NOTES D'UN AVIATEUR. Paris: 1918.

130. La Frégeolière, Renaud de. CROISIERES AERIENNES. SOUVENIRS D'UN PILOTE DE CHASSE. LE G.C II 1916-1918. Preface by Général Weygand. Paris: 1933.

131. ________. A TIR D'AILES. CARNET DE VOL D'UN AVIATEUR ET SOUVENIRS D'UN PRISONNIER. Paris: 1916.

132. Laine, André. POUR REUSSIR LES EPREUVES DU BREVET MILITAIRE, CONSEILS PRATIQUES. Paris: 1917.

133. Lainé, Serge. "L'Aéronautique militaire française au Maroc (1911-1939)." R.H.A. (No. 4, 1978), 107-119.

134. Langevin, Capitaine H. NUNGESSER. Preface by Capitaine Fonck. Paris: n.d.

135. Laurentie, J. LES LOIS DE LA GUERRE: GUERRE SUR TERRE, GUERRE AERIENNE, GUERRE MARITIME, SANCTION DES LOIS DE LA GUERRE, NEUTRALITE, FIN DE LA GUERRE. Paris, 1917.

136. Laux, James M. "Gnome et Rhône - an Aviation Firm in the First World War." A.H. 27 (Spring, 1980), 17-26.

137. Le Grand, G. COMMENT SERVIR DANS L'AERONAUTIQUE MILITAIRE. Preface by Colonel Girod. Paris: 1918.

138. Lemperiere, L. LE BOMBARDEBOCHES. PROJECT D'UN APPAREIL VISEUR POUR LE BOMBARDEMENT PAR AVIONS. Tours: 1915.

139. Lhermitte, Pierre. GUYNEMER 1917-1982. POELKAPELLE SE SOUVIENT. Orléans: 1982.

140. Liron, Jean. LES AVIONS FARMAN. Paris: 1984.

141. Lissarague, Général Pierre. MEMENTO TO D'HISTOIRE DE

L'AERONAUTIQUE MILITAIRE FRANCAISE: LA GUERRE DE 1914-1918. Salon de Province: 1978.

142. Lucas, Commandant Jean. "L'aviation de D.C.A. en 1917." R.A.A. 59 (June, 1934), 673-686.

143. ________. LA D.C.A. DE SES ORIGINES AU 11 NOVEMBRE 1918. Preface by Général Perrier. Paris: 1934.

144. ________. "La chasse de nuit pendant la guerre de 1914-1918." R.A.A. 69 (Apr., 1935), 383-404.

145. ________. "Emploi des ballons dahs la défense aérienne." R.A.A. 85 (Aug., 1936), 863-887.

146. McConnell, J.R. FLYING FOR FRANCE: WITH THE AMERICAN ESCADRILLE AT VERDUN. New York: 1918.

147. Marc, Lieutenant [Jean Beraud-Villars]. NOTES D'UN PILOTE DISPARU. Paris: 1918. See entry 26.

148. Marchal, Lieutenant Anselme. APRES MON VOL AU-DESSUS DE BERLIN, MA CAPTIVITE, MES EVASIONS. Paris: 1919.

149. Marellec, Fernand le. "La vie ardente de Guynemer." F.A.F. 176 (Dec., 1961), 783-803.

150. Mariage, Colonel Pierre. L'ADIEU AUX AILES. Paris: 1967.

151. ________. "L'adieu aux ailes. La MF 1 en Alsace." F.A.F. 217 (Aug.-Sept., 1965), 225-250.

152. ________. "L'adieu aux ailes. L'escadrille AR-272." F.A.F. 219 (Nov., 1965), 563-584.

153. ________. "L'adieu aux ailes. L'escadrille AR-272 en Alsace." F.A.F. 220 (Dec., 1965), 789-804.

154. ________. "L'adieu aux ailes: l'escadrille AR 272 à Verdun et l'armistice en Lorraine." F.A.F. 221 (Jan., 1966), 79-97.

155. Marie, Colonel F. "Comment la France a assuré son recrutement de pilotes aviateurs pendant la guerre de 1914-1918." R.A.M. 14 (Mar.-Apr.,

1923), 23-26; 15 (May-June, 1923), 49-51; 17 (Sept.-Oct., 1923), 106-111.

156. Marill, Jean-Marc. "1914-1918, l'Aéronautique militaire française. Naissance de la cinquième arme." Unpublished thesis, doctorat de IIIe cycle, Université de Paris I, 1985.

157. Martel, R. L'AVIATION FRANCAISE DE BOMBARDEMENT. DES ORIGINES AU 11 NOVEMBRE 1918. Preface by Général de Goÿs. Paris: 1939.

158. Mason, Herbert Molloy. HIGH FLEW THE FALCONS: THE FRENCH ACES OF WORLD WAR I. Philadelphia: 1965.

159. ________. THE LAFAYETTE ESCADRILLE. New York: 1964.

160. Massenet de Marancour, Commandant. LA CHASSE EN AVION. Paris: 1921.

161. Maurras, Charles. LES NUITS D'EPREUVE ET LA MEMOIRE DE L'ETAT. CHRONIQUE DU BOMBARDEMENT DE PARIS. Paris: 1923.

162. Melon, P. LE SANG DES AIGLES. Paris: 1931.

163. Meyer, Commandant André. AVOIR DES AILES. Preface by Général Stehlin. Nancy: 1967.

164. Milan, René. MATELOTS AERIENS 1916/1917. Paris: 1919.

165. ________. TROIS ETAPES: L'ARMEE D'ORIENT, L'AVIATION MILITAIRE, L'ITALIE. Paris: 1917.

166. ________. LES VAGABONDS DE LA GLOIRE. Paris: 1916.

167. Montalègre, Henri. LE LIEUTENANT BOURJADE: PILOTE ET MISSIONNAIRE, L'AS AUX 32 VICTOIRES. Paris: 1940.

167A. Morgat, Lieutenant Colonel L. "L'aviation en Berry pendant la Grande Guerre." R.H.A. (No. 1, 1980), 158-216.

168. Morizon, Alain. "L'aviation française en 1916." R.H.A. (No. 3, 1966), 40-52.

169. ________. "l'aviation en 1917." R.H.A. (No. 4, 1967),

21-36.

170. Mortane Jacques. LES AILES DE LA MORT (1914-1918-19??). Paris: 1932.

171. _________. LES AS NOUS PARLENT. Paris: 1936.

172. _________. L'AVIATION A TIR D'AILE. Tours: 1940.

173. _________. CARRE D'AS. Paris: 1947.

174. _________. LES DIRIGEABLES TRAGIQUES. Paris: 1938.

175. _________. EVASIONS D'AVIATEURS. Paris: 1928.

176. _________. LA GUERRE DES AILES (1914-1918). Paris: 1932.

177. _________. LA GUERRES DES NUAGES RACONTEE PAR SES MORTS. Paris: 1918.

178. _________. GUYNEMER, THE ACE OF ACES. New York: 1918.

179. _________. LES HEROS DE L'AIR. Paris: 1936.

180. _________. HISTOIRE ILLUSTREE DE LA GUERRE AERIENNE, 1914-1918. Paris: n.d.

181. _________. LEUR DERNIER VOL. Paris: 1931.

182. _________. MISSIONS SPECIALES. Paris: 1929.

183. _________. LES MYSTERES DE LA GUERRE AERIENNE. Paris: 1919.

184. _________. NAVARRE, SENTINELLE DE VERDUN. Paris: 1930.

185. _________. SENTINELLES DE L'AIR. Paris: 1936.

186. _________. SPECIAL MISSIONS OF THE AIR. AN EXPOSITION OF SOME OF THE MYSTERIES OF AERIAL WARFARE. London: 1919.

187. _________. TRAQUES PAR L'ENNEMI. Paris: 1929.

188. _________. A TRAVERS LES FILETS DE L'ENNEMI. Paris: 1929

189. ________. LES VOLS EMOUVANTS DE LA GUERRE. Paris: 1917.

190. Mortureux, André. DES FRANCAIS VIVENT L'EPOPEE SERBE 1914-1915. Paris: 1964.

191. "Opérations de la première division aérienne. Historique succinct établi d'après des documents officiels." R.A.M. 4 (July-Aug., 1921), 73-78.

192. Orthlieb, Commandant. L'AERONAUTIQUE. HIER, DEMAIN.... Paris: 1920.

193. Paquier, Général. "Le Général Louis de Goÿs, créateur du bombardement." F.A.F. 241 (Nov., 1967), 669-700.

194. Parsons, Edwin C. I FLEW WITH THE LAFAYETTE ESCADRILLE. Indianapolis: 1963.

195. Percheron, Maurice. NOS AVIONS. L'ESSOR ET L'ATTERRISSAGE. Paris: 1914.

196. ________. LES AEROPLANES DE LA GUERRE. LES AEROPLANES DE 1916 A 1920. Paris: 1920.

197. Perrin de Brichaimbaut, Capitaine. AVIATION DE GUERRE. Paris: 1923.

198. Pesquiès-Courbier, Simone. "L'aeronautique militaire française 1914-1918." ICARE 85 (1978), 3-161; 86 (1979), 5-119.

199. ________. "Le Groupe de Bombardement No. 1 durant la guerre de 1914-1918." R.H.A. (No. 3, 1983), 88-106.

200. ________. "La politique de bombardement des usines sidérurigiques en Lorraine et an Luxembourg pendant la première guerre mondiale." R.H.A. (No. 4, 1981), 126-159.

201. ________. "Les sidérurgistes de l'Est et la politique de bombardement d-usines en 1914-1918." R.H.M.C. (Jan.-Mar., 1984), 53-74.

202. Petit, Edmond. "L'aviation française en 1918." R.H.A.

(No. 4, 1968), 159-178.

203. ________. L'AVIATION PENDANT LA PREMIERE GUERRE MONDIALE. Paris: 1970.

204. ________. "Edouard Barrès." F.A.F. 255 (Feb., 1969), 179-196.

205. ________. "Histoire des forces aériennes françaises." ICARE 92 (1980), 3-85.

206. ________. LA VIE QUOTIDIENNE DANS L'AVIATION EN FRANCE AU DEBUT DU XXe SIECLE 1900-1935. Paris: 1977.

207. Poirier, Jules. LES BOMBARDEMENTS DE PARIS (1914-1918): AVIONS-GOTHAS-ZEPPELINS-BERTHAS. Paris: 1930.

208. Porret, Daniel. DICTIONNAIRE DES AS FRANCAIS DE LA GRANDE GUERRE. 2 vols. Vincennes: 1983.

209. Pujo, Bernard. "Evolution de la pensée du Général Foch sur l'emploi de l'arme aérienne an cours de son commandement du groupe des armées du Nord en 1915 et 1916." COLLOQUE INTERNATIONAL. ADAPTATION DE L'ARME AERIENNE AUX CONFLITS CONTEMPORAINS ET PROCESSUS D'INDEPENDANCE DES ARMEES DE L'AIR DES ORIGINES A LA FIN DE LA SECONDE GUERRE MONDIALE. Paris: 1985.

210. Rafex. "French Naval Airships. Their Development during the War." FLIGHT Oct. 2, 1919, 1315-1317.

211. Reinach, J. LES COMMENTAIRES DE POLYBE. SIXIEME SERIE. Paris: 1916.

212. Renaitour, Jean-Michel. LES FILS D'ICARE. HISTOIRE D'AVIATION. Paris: 1933.

213. Renard, Lieutenant Colonel P. L'EVOLUTION DE L'AERONAUTIQUE PENDANT LA GUERRE (1914-1918). Coulommier: 1921.

214. Reynolds, Quentin. THEY FOUGHT FOR THE SKY. THE DRAMATIC STORY OF THE FIRST WAR IN THE AIR. New York: 1957.

215. Robertson, Bruce. AIR ACES OF THE 1914-1918 WAR. Letchworth: 1959.

216. Rolland, Louis. LES PRATIQUES DE LA GUERRE AERIENNE DANS LE CONFLIT DE 1914 ET LE DROIT DES GENS. Paris: 1916.

217. Rougevin-Baville. "L'aëronautique militaire française, les débuts de la guerre aérienne: 1914." R.H.A. (No. 3, 1964), 108-118.

218. Sahel, Jacques. HENRY FARMAN ET L'AVIATION. Paris: 1936.

219. Segretain, Colonel. "L'aviation d'observation française au début de la guerre 1914-1918." F.A.F. 143 (Dec., 1958), 849-872.

220. Sensever, Lieutenant G. LE COMBAT AERIEN. Paris: 1917.

221. Shaffer, Walter J. "Do it Yourself Flying.... One Man's Recollection of the Lafayette Escadrille." A.H. (Summer, 1970), 83-87.

222. Sizaire, P. "Maurice Arnoux." F.A.F. 9 (June, 1947), 293-310.

223. Tenant de la Tour, Geoffroy. EN SOUVENIR DES CIGOGNES. Preface by Commandant Brocard. Limoges: 1921.

224. Suarnet, André. LES RAMPANTS, SOUVENIRS D'ESCADRILLE. Paris: 1930.

225. Tessan, F. de. QUAND ON SE BAT. Paris: 1916.

226. Thébault, Louis. L'ESCADRILLE 210. Paris: 1925.

227. Thénault, Georges, Lieutenant-Colonel. L'ESCADRILLE LAFAYETTE. (AVRIL 1916-JANVIER 1918). Preface by Général Gouraud. Paris: 1939.

228. Thiéry, Maurice. PARIS BOMBARDE PAR ZEPPELINS, GOTHAS ET BERTHAS. Paris: 1921.

229. Thomas, M.R. "Les premiers raids aériens en Afrique 1911-1933." F.A.F. 163 (Dec., 1960), 739-768.

230. Thoret, J. "Pilotes d'il y a vingt-cinq ans." R.A.A. 82 (May, 1936), 496-500.

231. "Training Airmen in France." LITERARY DIGEST Nov. 18, 1916, 1324-1325.

232. Verdurand, Capitaine. COMMENT OBTIENDRA-T-ON LE RENDEMENT MAXIMUN DE L'OBSERVATION AERIENNE ET D'ARTILLERIE. Paris: 1918.

233. Verneuil, Max. AEROSTIER, MON CAMARADE... Preface by Général Hirschauer. Paris: 1934.

234. Vigilant. FRENCH WAR BIRDS. London: 1937.

235. Vilain, C. L'ESCADRILLE DES SIOUX. Paris: 1940.

236. Vincent, Daniel. LA BATAILLE DE L'AIR. Paris: 1918.

237. Violan, Jean. DANS L'AIR ET DANS LA BOUE. MES MISSIONS DE GUERRE. Paris: 1933.

238. Voisin, Général. LA DOCTRINE DE L'AVIATION FRANCAISE DE COMBAT AU COURS DE LA GUERRE (1915-1918). Paris: 1932.

239. _________. "La doctrine de l'aviation française de combat en 1918." R.F.A. 25 (Aug., 1931), 895-914; 26 (Sept., 1931), 1029-1058; 27 (Oct., 1931), 1151-1177; 28 (Nov., 1931), 1271-1301.

240. Weiller, Paul-Louis. "René Fonck (1894-1953)." REVUE DES DEUX MONDES (No. 6, 1983), 533-540.

241. Weiss, Pierre. LA GUERRE AERIENNE DEVANT NANCY. Nancy: 1921.

242. _________. LES CHARMEUR DE NAUGES. Paris: 1924.

243. _________. PINSARD CHEZ LES BOCHES. Paris: 1935.

244. Winslow, Carroll Dana. WITH THE FRENCH FLYING CORPS. New York: 1917.

THE INTERWAR PERIOD

Basic general studies on the period between the wars are still lacking, but numerous dissertations under way promise to fill a number of gaps in our knowledge. We could mention in this regard the thesis being prepared by M. Hodeir on aviation the period of the Armée de l'Air's struggle for independence from 1919 to 1933, the thesis being prepared by D. Hary on air rearmament from 1940, and that being done by J. Blanc on French fighter aviation from 1919 to 1939. A Special number of the REVUE HISTORIQUE DES ARMEES published in 1977 was devoted to French military aviation and is an excellent point of departure; the great variety of subjects touched upon could well inspire a host of new studies.

The problems in fixing air doctrine for the twenty-year period following World War I can best be approached through an excellent article which Colonel LeGoyet published in 1969. Clear and precise, this work lays out the basic problems. Also useful is an article by P. Buffotot on air doctrine in the interwar period (518) and a communication by General P.M. Gallois on French air doctrine in 1939 (266). The influence of the ideas of Giulio Douhet in France may be traced in three works by Lieutenant-Colonel Vauthier (777, 778, 780). A study by Maurice Vaïsse on bombardment aviation (771) shows the problems of disarmament and also the moral opprobrium which weighed upon the bomber force in the early thirties. There is also a certain number of studies which show the Armée de l'Air in some of the crises of the thirties, for example the work of Patrice Buffotot and General Charles Christienne on the air force and the crisis of March 7, 1936 (126), Patrick Facon's study of the air aspects of the Munich Crisis (246), and Patrice Buffotot's study of German rearmament as perceived by the French intelligence agencies from 1936 to 1939 (123). The problem of remarmament in the air can be better understood in its financial aspects through the much cited work of R. Frankenstein (261).

General Armengaud's book on the Rif War (27) remains the best written to date on the use of aviation in essentially colonial campaigns (the author commanded the air forces in that war).

In quite another area of inquiry, the relations between the national legislature and the air force for the period 1928-1940 have been treated in depth and precision by D. Boussard (106) in a book which remains the fundamental work on the subject. One might add M. Hodeir's study of aviation and the "Horizon Blue Chamber" (337). In this period, as for preceding ones, the thesis of E. Chadeau, cited earlier, is the key to all study of the aircraft industry. Chadeau's work could nevertheless completed by his study of the interplay of tactical doctrine and the evolution of the aircraft industry from 1912 to 1928 (143).

The Franco-British Alliance and the cooperation in air matters which developed between the two countries from 1935 on have been studied exhaustively in an excellent book of P. Fridenson and J. Lecruir (265). French aircraft purchases in the United States have been covered in the work of J. Haight (298-302).

The subject of attitudes and perceptions has been taken up by Y. Brèche in a study of air force officers between the wars (108). Finally, Patrick Facon has written an article on popular aviation which offers clues to problems of recruitment during the period (244).

BIBLIOGRAPHICAL LISTING

1. A..., Général. "La puissance offensive de l'instrument de guerre de demain." R.F.A. 37 (Aug., 1932), 343-392; 38 (Sept., 1932), 965-1003; 39 (Oct., 1932), 1079-1105; 40 (Nov., 1932), 1199-1226.

2. A.B. "La guerre totale et l'aviation." L.A. 536 (Sept. 24, 1931), 5.

3. Accart, Jean. CHASSEUR DU CIEL. HISTOIRE DE LA LERE ESCADRILLE DU GROUPE DE CHASSE 1/5. Preface by Général Blanchard. Paris: 1941.

4. ________. ON S'EST BATTU DANS LE CIEL. Grenoble: 1942.

5. "L'activité aérienne militaire. Travail aérien dans les formations. Voyages en France, hors de France." R.A.M. 42 (Nov.-Dec., 1927), 121.

6. "L'activité de l'aviation militaire française." R.A.M. 11 (Sept.-Oct., 1922).

7. "L'aéronautique maritime." R.A.M. 12 (Nov.-Dec., 1922), 125-127.

8. "L'aéronautique maritime en France." R.F.A. 20 (Mar., 1931), 347; 21 (April, 1931), 385-438.

9. "L'aéronautique militaire en Indochine. Mission 'Sud-Annam'." R.A.M. 31 (Jan.-Feb., 1926), 5-15.

10. "L'aéronautique militaire et les progrès de la construction aéronautique." R.A.M. 24 (Nov.-Dec., 1924), 121-128.

11. "L'aéronautique au Maroc. Réduction de la tâche de Taza." R.A.M. 37 (Jan.-Feb., 1927), 16-19.

12. ALERTE AUX AVIONS. MANUEL OFFICIEL ELEMENTAIRE DE DEFENSE PASSIVE CONTRE LES ATTAQUES AERIENNES. Paris: 1938.

13. Alix, Roger. "La lutte de l'avion et du canon." L.A. 525 (Sept. 7, 1931), 5.

14. Alléhaut, Général. ETRE PRETS. PUISSANCE AERIENNE. FORCES DE TERRE. DOCTRINE. ORGANISATION. MORALE. Paris: 1935.

15. Andral, Paul. "L'Air dans la défense nationale." L.A. 554 (Jan. 28, 1932), 5.

16. ________. "L'aviation au Parlement. La loi d'organisation générale." L.A. 545 (Dec. 17, 1931), 5.

17. ________. "L'aviation en face de l'Armée." L.A. 561 (March 3, 1932), 5-6.

18. ________. "La crise: les huit triumvirats." L.A. 560 (March 19, 1932), 5.

19. ________. "La défense aérienne." L.A. 549 (Dec. 24, 1931), 5.

20. ________. "Douhet et la 'guerre de l'Air'." L.A. 552 (Jan. 14, 1932), 5.

21. ________. "Douhet et l'Etat-Major français." L.A. 573 (June 9, 1932), 5.

22. ________. "Le mal et son remède." L.A. 557 (Feb. 18, 1932), 5.

23. ________. "Union et silence." L.A. 559 (March 3, 1932), 5.

24. Angeli, Jean. "La vitesse du chasseur et du bombardier." L.A. 736 (July 25, 1935), 6.

25. Anget, E. and R. de Lavergne. UNE FIGURE LEGENDAIRE DE L'AVIATION FRANCAISE DE 1914 A 1940. LE GENERAL VUILLEMIN. LE COMBATTANT. LE PIONNIER DU SAHARA. LE CHEF. Paris: 1965.

26. "L'Armée de l'Air." FLUGWEHR UND-TECHNIK I (No. 3, 1939), 61-65.

27. Armengaud, Général. QUELQUES ENSEIGNEMENTS DES CAMPAGNES DU RIFF EN MATIERE D'AVIATION (1925-1926). Paris: 1928.

28. Astorkia, Madeline. "Les lecons aériennes de la guerre d'Espagne." R.H.A. (No. 2, 1977), 145-175.

29. Astruc, Capitaine. "L'expérience actuelle. Instruction des observateurs en avion." R.A.M. 47 (Sept.-Oct., 1928), 113-117; 48 (Nov.-Dec., 1928), 135-140.

30. Astruc, Commandant. "Les réserves générales d'aviation." R.F.A. 46 (May, 1933), 483-507.

31. "L'attaque des bases aériennes." R.A.A. 108 (July, 1938), 723-725.

32. Auber, A.J. CAPITAINE AU GROUPE DE CHASSE "NAVARRE" (1918-1944). Paris: 1946.

33. Auphan, Capitaine. "L'aviation militaire en Indochine." R.F.A. 30 (Jan., 1932), 3-11.

34. "L'aviation coloniale." R.A.M. 12 (Nov.-Dec., 1922), 138-142.

35. "L'aviation dans une opération de débarquement." R.A.A. 95 (June, 1937), 603-605.

36. "L'aviation et la défense nationale." L.A. 292 (Dec. 20, 1928), 10-11.

37. "L'aviation militaire et la pénétration saharienne." R.A.M. 28 (July-Aug., 1925), 73-84.

38. "L'aviation militaire au Maroc au cours du deuxième trimestre 1925." R.A.M. 29 (Sept.-Oct., 1925), 47-112; 30 (Nov.-Dec., 1925), 121-125.

39. Ba, P. "Considérations techniques sur le tir de l'avion-canon." L.A. 652 (Dec. 14, 1933), 13.

40. ________. "Le monoplace-canon peut-il interdire l'action offensive de l'aviation lourde?" L.A. 675 (May 24, 1934), 4.

41. ________. "Le multiplace de combat est-il un réalité militaire?" L.A. 646 (Nov. 2, 1933), 12.

42. ________. "La tactique et l'armement des avions de

chasse modernes." L.A. 694 (Oct. 4, 1934), 11.

43. Barjot, Pierre. L'AVIATION MILITAIRE FRANCAISE. Preface by Général Vuillemin. Paris: 1937.

44. ________. "L'avion de bataille de Douhet est-il un archaïsme?" R.F.A. 53 (Dec., 1933), 1323-1332.

45. ________. "Chasse de nuit et chasse de jour." R.A.A. 66 (Jan., 1935), 27-36.

46. ________. "Chasse de nuit: monoplace ou biplace?" R.A.A. 71 (June, 1935), 623-632.

47. ________. "Douhet et la technique." R.M.A. 13 (Jan., 1936), 14-18.

48. ________. "Emploi de l'aviation dans la guerre navale." R.F.A. 27 (Oct., 1931), 1217-1223.

49. ________. "La guerre aux bases dans la guerre aérienne." R.A.A. 63 (Oct., 1933), 1139-1162.

50. ________. "La guerre aux bases dans la guerre maritime." R.F.A. 48 (July, 1933), 723-733.

51. ________. "Moteur-canon, le canon léger ou multitube lourd." R.A.A. 68 (March, 1935), 263-272.

52. ________. "La neutralité du ciel." R.A.A. 94 (May, 1937), 500-503.

53. ________. "A propos de l'offensive aérienne dans la guerre sur mer." R.F.A. 42 (Jan., 1933), 60-90.

54. Barthélémy, Commandant. "L'aviation dans l'Afrique du Nord." R.A.M. 1 (Jan.-Feb., 1921), 7-13.

55. ________. "L'aviation et les chars de combat." R.A.M. 2 (March-April, 1921), 22-31.

56. Barthélemy, Général. L'ESCADRILLE DU CALAO. Preface by Général F. Maurin. Paris: 1976.

57. Barzini, Luigi. LA GUERRE MODERNE, SUR TERRE, DANS LES AIRS ET SOUS LES EAUX. Paris: n.d.

58. Baudry de Saunier. COMMENT PARIS A ETE DETRUIT EN SIX HEURES. Paris: 1921.

59. Baume, Commandant de la. LES MOYENS DE L'AERONAUTIQUE DE CORPS D'ARMEE. LEUR MISE EN OEUVRE DANS LA BATAILLE ET DANS LES EXERCICES DU TEMPS DE PAIX. Paris: 1929.

60. Beaufort de. "Deux opinions intéressantes sur le monoplace-canon." L.A. 638 (Sept. 7, 1933), 5.

61. Bechoff, Roland. L'AVIATION ET LE DESARMEMENT. Paris: 1932.

62. Belleroche, Pierre. HISTOIRE DU COMBAT AERIEN. Marseille: 1941.

63. Bénazet, Paul. NOTRE SECURITE. Paris: 1938.

64. Benoist, Général. L'AVIATION DE MONTAGNE. Grenoble: 1934.

65. Benoist, Marc. L'AVIATION FRANCAISE. Paris: 1936.

66. Berande, Charles. "L'aviation dans la politique internationale." L.A. 589 (Sept. 29, 1932), 5.

67. _________. "Controverse sur le croiseur aérien." L.A. 526 (July 17, 1931), 5; 527 (July 23, 1931), 5; 531 (Aug. 20, 1931), 5; 533 (Sept. 3, 1931), 5.

68. _________. "Les dangers d'une fausse théorie." L.A. 641 (Sept. 28, 1933), 13.

69. _________. "La défense active? Une couteuse illusion!" L.A. 569 (May 12, 1932), 5.

70. _________. "La défense aérienne du territoire. Ce que nous faisons; ce que font les autres." L.A. 676 (May 31, 1934), 13.

71. _________. "Le désarmement aérien." L.A. 546 (Dec. 3, 1931), 5.

72. _________. "Le désarmement aérien. Genève et la limitation des armements aèriens." L.A. 547 (Dec. 10, 1931), 5, 6.

73. ________. "Le désarmement aérien. Pas un avion! Pas un pilote!" L.A. 553 (Jan. 21, 1932), 5.

74. ________. "Le désarmement aérien. La thèse de la France." L.A. 550 (Dec. 31, 1931), 5.

75. ________. "Le désarmement aérien. Les travaux de la commission préparatoire." L.A. 549 (Dec. 24, 1931), 5, 6.

76. ________. "Le désarmement aérien. Les trois dangers." L.A. 555 (Feb. 4, 1932), 5.

77. ________. "Esquisse d'une nouvelle tactique de la lutte aérienne. L'influence du monoplace-canon sur la conception du bombardement, de la reconnaissance et de l'observation." L.A. 630 (July 13, 1933), 13.

78. ________. "Le monoplace-canon contre le multiplace de combat." L.A. 629 (July 7, 1933), 13.

79. ________. "Pas un escadron, mais toute l'aviation!" L.A. 556 (Feb. 1, 1932), 5.

80. ________. "Pour un conseil supérieur de l'Air." L.A. 523 (June 25, 1931), 5.

81. ________. "Les sept péchés capitaux." L.A. 561 (March 17, 1932), 5.

82. ________. "Sur les centres d'expériences." L.A. 522 (June 18, 1931), 5.

83. Bernard, Lieutenant de Vaisseau. "Contribution à l'étude d'une liaison aérienne France-Madagascar." R.F.A. 3 (Oct., 1929), 343-350.

84. Bienvenue, Colonel. HISTORIQUE DE L'AEROSTATION. Paris: 1934.

85. ________. "Faisons le point en aérostation." R.A.A. 61 (Aug., 1934), 883-904.

86. Beyne, Lieutenant Colonel. "A propos de la sélection

physique du personnel navigant. Quelques précisions." R.A.M. 46 (July-Aug., 1928), 42-45.

87. Blanc, Edouard. "Pelletier-Doisy." L.A. 1427 (June 6, 1953), 4.

88. Blériot, Louis, and Edward Ramon. LA GLOIRE, DES AILES. L'AVIATION DE CLEMENT ADER A COSTES. Paris: 1927.

89. Blery, Paul. "Aviation de coopération ou aviation d'assaut." L.A. 710 (Jan. 24, 1935), 11.

90. Bloch, Lieutenant Colonel. LA GUERRE CHIMIQUE... Preface by Général Weygand. Paris: 1927.

91. Boisboissel, Lieutenant de. "L'avion, moyen de commandement aux colonies." R.F.A. 36 (July, 1932), 779-785.

92. "Le bombardement des services de l'arrière." R.A.A. 87 (Oct., 1936), 1083-1844.

93. LE BOMBARDEMENT DES VILLES OUVERTES. Paris: 1938.

94. Boniface, Lieutenant. "Le personnel de l'aviation." R.A.M. 6 (Nov.-Dec., 1921), 130-132.

95. Boucabeille, Général. "L'armée de l'Air." L.A. 266 (July 22, 1926), 1.

96. ________. "Notre ciel est-il défendu? Défense nationale et armée de l'Air." L.A. 482 (Sept. 11, 1930), 1, 2.

97. ________. "L'aviation en Indochine." R.F.A. 18 (Jan., 1931), 28-37.

98. Bouché, Henri. "La force nationale et la résistance à l'agression de surprise." R.A.A. 108 (July, 1938), 726-736.

99. Bouché, H. and P. Valentin. "Pour une force aérienne internationale de contrôle et d'appoint." R.A.A. 96 (July, 1937), 825-832.

100. Bourdès, Colonel. "La création de l'arme aéroportée

française." R.H.A. (No. 2, 1977), 62-86.

101. ________. "Quand les parachutistes portaient la casquette bleue." A.A. 302 (Sept.-Oct., 1976), 50-59.

102. Bourdet, Maurice. GRANDEUR ET SERVITUDE DE L'AVIATION. Paris: n.d.

103. Bourg, Charles. "Les conditions d'une limitation de l'emploi de l'aviation." L.A. 897 (Aug. 23, 1938), 11.

104. Bouruet-Aubertot, J. LES BOMBARDEMENTS AERIENS. Paris: 1928.

105. Bouscat, Commandant. "Une mission aérienne dans la région du Tibesti." R.F.A. 11 (June, 1930), 709-713.

106. Boussard, Dominique. UN PROBLEME DE DEFENSE NATIONALE: L'AERONAUTIQUE AU PARLEMENT (1928-1940). Vincennes: 1985.

107. Boutiron, P. "L'aviation navale et son rôle dans la marine de guerre." R.A.M. 22 (July-Aug., 1924), 73-80; 23 (Sept.-Oct., 1924), 97-101.

108. Brèche, Yves. "Les officiers de l'armée de l'Air à travers la littérature militaire." R.H.A. (No. 2, 1977), 113-144.

109. Bréguet, Louis. "Les fabrications aéronautiques et la chambre syndicale." L.A. 76 (Nov. 30, 1922), 1.

110. ________. "Réflexions sur l'évolution technique des avions." R.A.A. 81 (April, 1936), 394-410.

111. Brereton, L.H. "France, Master of the Air." U.S. AIR SERVICE 8 (Feb., 1923), 7-11.

112. Breyton, W. Capitaine. "L'aviation du Maroc et l'affaire du Sagho." R.A.A. 63 (Oct., 1934), 1107-1138.

113. ________. "L'aviation sanitaire au Maroc en 1933." R.A.A. 56 (Mar., 1933), 243-264.

114. ________. "L'aviation et la pacification du grand Atlas central." R.A.A. 64 (Nov., 1934), 1265-1310.

115. Brion, Maurice. "Une controverse. Les problèmes de la chasse." L.A. 447 (Jan. 9, 1930), 5, 6.

116. ________. "Faut-il placer des canons sur tous les avions militaires?" L.A. 636 (Aug. 24, 1933), 11.

117. Brissot, Ingénieur. "Artillerie de bord des croiseurs et des contre croiseurs aériens." R.A.A. 101 (March, 1938), 269-284.

118. ________. "Probabilité et cadence du tir d'avion contre canon." R.A.A. 100 (Nov., 1937), 1257-1278.

119. Bru, Commandant. L'AERONAUTIQUE. COURS D'EMPLOI DES ARMES. Saint Maixent: 1924.

120. Bülow, Freiherr von. "Die Grossluftmacht Frankreichs." WISSEN UND WEHR 14 (1933), 225-232.

121. Buffotot, Patrice. "La doctrine aérienne du commandement français pendant l'entre deux guerres." REVUE INTERNATIONALE D'HISTOIRE MILITAIRE 55 (1983), 169-184.

122. ________. "La doctrine aérienne du haut commandement pendant l'entre deux guerres." ARMEE DEFENSE Sept.-Oct., 1980, 31-36.

123. ________. "La perception du réarmement allemand par les organismes du renseignement français de 1936 à 1939." R.H.A. (No. 3, 1979), 173-184.

124. ________. "Le réarmement aérien allemand et l'approche de la guerre vus par le II° Bureau Air français (1936-1939)." DEUTSCHLAND UND FRANKREICH 1936-1939. DEUTSCH-FRANZOESISCHE HISTORIKER COLLOQUIUM. Bonn: 1982, 249-289, 648-649.

125. Buffotot, Patrice and Général Charles Chistienne. "L'aéronautique militaire française entre 1919 et 1939." R.H.A. (No. 2, 1977), 9-40.

126. ________. "L'Armée de l'Air française et la crise du 7 mars 1936." FRANCE ET ALLEMAGNE (1932-1936). Paris: 1980, 315-331.

127. Bui Xuan Bao. AVIATION ET LITTERATURE: NAISSANCE D'UN HEROISME NOUVEAU DANS LES LETTRES FRANCAISES DE L'ENTRE DEUX GUERRES. Paris: 1960.

128. C.M. "Réflexions sur l'aviation d'artillerie." R.A.A. 113 (Dec., 1938), 1331-1342.

129. Campardon, Lieutenant de Vaisseau. "L'escadrille 5 B2 au Maroc." R.F.A. 2 (Sept., 1929), 131-155.

130. Canonne, Capitaine. P. "L'aviation d'infanterie." R.A.M. 8 (Mar.-Apr., 1922), 25-27; 10 (July-Aug., 1922), 77-82.

131. ________. "L'aviation et la cavalerie." R.A.M. 14 (Mar.-Apr., 1923), 35-37.

132. Cantener, Colonel Henri. HOMMAGE AUX AILES LORRAINES. Preface by Général M. Valin et Général L. Delfine. Nancy: 1968.

133. Caquot, A. "Maintenant plus que jamais notre aviation a besoin d'un vigoureux effort technique." L.A. 3/2 (Aug. 3, 1928), 9, 10.

134. Carlier, Claude. "L'évolution des doctrines de guerre aérienne à travers l'enseignement dispensé à l'Ecole Supérieure le guerre de 1918 à 1928." COLLOQUE INTERNATIONAL SUR L'ADAPTATION DE L'ARME AERIENNE AUX CONFLITS CONTEMPORAINS ET PROCESSUS D'INDEPENDANCE DES ARMEES DE L'AIR. Paris: 1985, 81-96.

135. Casse, Lieutenant-Colonel. "Faut-il des 'divisions aériennes'?" R.A.M. 5 (Sept.-Oct., 1921), 97-99.

136. Castellane, Jean. "Hésitations injustifiées." L.A. 326 (Sept. 15, 1927), 3.

137. Castex, Marill. ETUDE PHYSIO-PATHOLIGIQUE DE L'AVIATION DE GUERRE. Toulouse: 1921.

138. Castillon de Saint-Victor, A. de. "Paris-Tokio par

Pelletier-Doisy." F.A.F. 242 (Dec., 1967), 813-842.

139. Césari, Lieutenant-Colonel. "L'encadrement des formations volantes." R.A.M. 97 (Aug., 1937), 889-902.

140. "C'est dans l'air qu'est la menace. Les Ailes au Parlement." L.A. 299 (Mar. 10, 1927), 1, 2.

141. Chabannes, H. de. PORT DE L'AIR. Paris: 1938.

142. Chadeau, Emmanuel. "L'Etat et les industries de construction aéronautique en France des débuts de l'aviation au réarmement (1900-1935)." Unpublished thesis, doctorat de troisième cycle, Université de Paris X Nanterre, 1979.

143. _________. "Un jeu d'influences réciproques. Doctrine d'emploi tactique et evolution industrielle aéronautique. Le cas français (1912-1928)." COLLOQUE INTERNATIONAL SUR L'ADAPTATION DE L'ARME AERIENNE AUX CONFLITS CONTEMPORAINS ET PROCESSUS D'INDEPENDANCE DES ARMEES DE L'AIR. Paris: 1985, 239-249.

144. _________. "Laurent Eynac et l'aéronautique (1920-1940)." CAHIERS HAUTE LOIRE, 1982, 195-210.

145. Chambe, René. ENLEVEZ LES CALES. Paris: 1935.

146. Champion, Capitaine Georges. "L'école militaire de l'air." F.A.F. 199 (Jan., 1964), 21-32.

147. Charpentier, J. L'HUMANISATION DE LA GUERRE AERIENNE. Preface by A. Geouffre de la Pradille. Paris: 1938.

148. "La chasse reste ouverte, plus que jamais." R.A.A. 88 (Nov., 1936), 1202-1220.

149. Chassey, Général X. de. "Le maréchal Fayolle, inspecteur général de l'aéronautique." F.A.F. 70 (July, 1952), 477-485.

150. Chassin, Général L.M. "Débuts de l'aviation en Indochine." F.A.F. 110 (Dec., 1955), 1003-1020.

151. Chateauvieux, Capitaine. "Doctrine d'organisation et d'emploi de l'aéronautique dans l'armée allemande." R.A.M. 34 (July-Aug., 1926), 73-78.

152. Cheutin, Commandant. "Role colonial de l'aviation militaire au Maroc." R.A.M. 4 (July-Aug., 1921), 85-91.

153. ________. "L'aviation française au Maroc." R.A.M. 16 (July-Aug., 1923), 77-81; 17 (Sept.-Oct., 1923), 100-104; 19 (Jan.-Feb., 1924), 1-6; 20 (Mar.-Apr., 1924), 25-30; 21 (May-June, 1924), 49-56.

154. Chevennes, François. LA GLOIRE A DES AILES. Preface by L. Bréguet. Paris: 1939.

155. Chollat, Henry. AVIATION POPULAIRE? AVIATION PRIVEE? AVIATION FRANCAISE? Preface by Laurent-Eynac. Paris: 1937.

156. Christienne, Général Charles. "Ader et Mitchell, les penseurs aéronautiques." R.H.A. (No. 1, 1982), 24-41.

157. Clermont-Tonnerre, F. de. "On veut démembrer le ministère de l'Air!" L.A. 888 (June 23, 1938), 11.

158. Cochet, Capitaine. "A propos de l'aviation d'infanterie." R.A.M. 12 (Nov.-Dec., 1922), 147-149.

159. Coint-Bavarot, Capitaine. "Les répercussions des progrès réalisés dans la construction aéronautique sur l'emploi des forces aériennes." R.F.A. 21 (Apr., 1931), 468-492.

160. ________. "Emploi comparé des canons et des mitrailleuses dans la bataille aérienne." R.A.A. 55 (Feb., 1934), 123-140.

161. ________. "Esquisse d'un programme de matériel d'aviation militaire." R.A.A. 63 (Oct., 1934), 1163-1188.

162. Un colonial. "Lettre d'Indo-Chine." L.A. 801 (Dec. 28,

1922), 4.

163. "Comment organiser l'aviation?" R.A.M. 13 (Jan.-Feb., 1922), 1-6.

164. Comoz, Pierre. "Une formule pratique du ministère de l'Air. La politique aérienne de la France." L.A. 362 (May 24, 1928), 5.

165. Contamin, L. and P. Fribourg. CE QU'IL FAUT SAVOIR DE L'AVIATION MILITAIRE. Paris: 1935.

166. Coox, Alvin D. "French Military Doctrine, 1919-1939; Concepts of Ground and Aerial Warfare." Unpublished doctoral dissertation, Harvard University, 1951.

167. Corbin, Louis. "Le recrutement actuel des officiers." L.A. 398 (Jan. 31, 1929), 5.

168. ________. "Il faut une école de l'Air." L.A. 401 (Feb. 21, 1929), 7, 8.

169. ________. "L'école des officiers de l'Air. Où et quand faut-il la créer?" L.A. 408 (Apr. 11, 1929), 6, 7.

170. ________. "L'école de l'Air. Nous demandons des chefs." L.A. 410 (May 25, 1929), 5.

171. ________. "L'école de l'Air sera." L.A. 447 (Jan. 9, 1930), 5.

172. ________. "A propos de l'école de l'Air." L.A. 528 (July 30, 1931), 5.

173. ________. "Pour l'école de l'Air 'intégrale'." L.A. 533 (Sept. 3, 1931), 6, 7.

174. ________. "L'école de l'Air. Quelques funestes erreurs." L.A. 541 (Oct. 29, 1931), 5.

175. ________. "L'école des sous-officiers navigants." L.A. 547 (Dec. 10, 1931), 5.

176. ________. "Les centres d'expériences tactiques et la

doctrine d'emploi de l'aviation." L.A. 551 (Jan. 7, 1932), 5.

177. ________. "Le front expérimental." L.A. 562 (Mar. 24, 1932), 5.

178. Cornillon, Capitaine and Lieutenant Gérardot. "Voyage Paris-Beyrouth et retour." R.A.M. 42 (Nov.-Dec., 1927), 123-126.

179. Cot, Pierre. "La nationalisation de l'industrie aéronautique. Discours pronocé par M. Pierre Cot, Ministre de l'Air, le 20 juillet 1937." R.A.A. 98 (Sept., 1937), 1018-1024.

180. ________. L'ARMEE DE L'AIR 1936-1938. Paris: 1939.

181. ________. LE PROCES DE LA REPUBLIQUE. New York: 1944.

182. ________. "The Defeat of the French Air Force." FOREIGN AFFAIRS 19 (No. 7, 1944), 790-805.

183. Cournot, Capitaine J. "Sur l'aviation d'observation et de reconnaissance." R.A.M. 5 (Sept.-Oct., 1921), 107-118.

184. Crochu, Commandant. "Considérations sur le bombardement aérien." R.F.A. 36 (July, 1932), 723-770.

185. Cuenat, Pierre. LA GUERRE AERO-CHIMIQUE. Preface by P. Cot. Paris: 1936.

186. Cuny, Jean and Raymond Danel. L'AVIATION DE CHASSE FRANCAISE 1918-1940. Paris: 1974.

187. ________. LE DEWOITINE D 520. Paris: 1975.

188. ________. L'AVIATION FRANCAISE DE BOMBARDEMENT ET DE RENSEIGNEMENT 1918-1940. Preface by Général P. Lissarague. Paris: 1980.

189. D., Lieutenant A. "Civils et combattants espagnols sous les bombardements aériens." L.A. 926 (Mar. 16, 1939), 11.

190. ________. "L'efficacité des bombardements én

Espagne." L.A. 927 (Mar. 23, 1939), 11.

191. Dagnaux, Capitaine. "L'aviation de bombardement. Son rôle, ses moyens, son état actuel." R.A.M. 4 (July-Aug., 1921), 79-84.

192. Dagnaux, Chef de Bataillon. "La liaison aérienne France-Madagascar. L'aviation à Madagascar." R.A.M. 41 (Sept.-Oct., 1927), 97-102.

193. Danel, Raymond. EMILE DEWOITINE. Paris: 1982.

194. "Le danger aérien est-il réel? L'opinion du Général Hirschauer." L.A. 341 (Dec. 29, 1927), 1, 2.

195. Déat, Marcel and Claude Bonnier. POUR UNE POLITIQUE DE L'AIR. Paris: 1938.

196. "La défense aérienne des 'Empires'." R.A.A. 103 (Feb., 1938), 123-124.

197. Delanney, Lieutenant-Colonel. "L'attaque aérienne massive du territoire." R.F.A. 10 (May, 1930), 499-523.

198. Denain, Commandant. "L'aviation au Levant de 1919 à 1923." R.A.M. 12 (Nov.-Dec., 1922), 129-134.

199. Demaison, André. MENACES DANS LE CIEL. Paris: 1933.

200. Desbordes, Pierre. "La maîtrise de l'Air. Le formidable programme britannique." L.A. 92 (Mar. 22, 1923), 1, 4.

201. ________. "Le danger aérien allemand. Les erreurs de prise en gage." L.A. 125 (Nov. 8, 1923), 3; 126 (Nov. 15, 1923), 3.

202. ________. "L'attaque et la défense du Bourget." L.A. 690 (Sept. 6, 1934), 9.

203. Detroyat, Michel. TU SERAS AVIATEUR. Paris: 1938.

204. Deve, Colonel M. "Le General Pelletier-Doisy." F.A.F. 85 (Oct., 1953), 271-278.

205. Devillers, Michel. "L'aviateur dans la mythologie

cinématographique." R.H.A. (No. 2, 1977), 195-204.

206. "Les discussions sur la question du bombardement aérien à la Conférence de La Haye (décembre 1922-février 1923)." R.A.M. 17 (Sept.-Oct., 1923), 93-99.

207. "Documents sur la limitation des armaments aériens." R.A.A. 113 (Dec., 1938), 1379-1408; 114 (Jan.-Feb., 1939), 49-76; 116 (May-June, 1939), 291-306.

208. Dollfus, Charles and Henri Bouché. HISTOIRE DE L'AERONAUTIQUE. Paris: 1938.

209. Doret, Marcel. TRAIT D'UNION AVEC LE CIEL. Paris: 1954.

210. "La dotation organique en aviation de coopération." R.A.A. 106 (May, 1938), 483-488.

211. Dumas, General E. "Les besoins de l'artillerie en aviation de coopération." R.A.A. 106 (May, 1938), 505-526.

212. Dumesnil, H. "Les pilotes militaires." L.A. 5 (July 21, 1921), 1.

213. Du Plessis, Jean. LES GRANDS DIRIGEABLES DANS LA PAIX ET LA GUERRE. 2 vols. Paris: 1925-26.

214. Duval, General. LES LECONS DE LA GUERRE D'ESPAGNE. Paris: 1938.

215. "L'école du vol rasant." R.A.A. 93 (Apr., 1937), 363-364.

216. "L'école militaire et d'application de l'Aéronautique à Versailles." R.F.A. 4 (Nov., 1929), 403-414.

217. "Eléments pour un historique de la coopération. I-L'aviation de coopération de 1918 à 1936. II-L'effort de l'armée de l'Air depuis 1936 dans l'action au profit des armées terrestres." R.A.A. 106 (May, 1938), 491-504.

218. "Emploi de l'aviation au Maroc en 1925." R.A.M. 32 (Mar.-Apr., 1926), 34-42; 33 (May-June, 1926), 57-62; 34 (July-Aug., 1926), 79-88.

219. Endres, F.C. LA GUERRE DES GAZ. Paris: 1928.

220. "En face des grandes sociétés nationales quelles firmes libres subsisterient?" L.A. 815 (Jan. 28, 1937), 11.

221. Escudier, Commandant. "L'aviation d'observation dans la guerre de mouvement." R.A.M. 35 (Sept.-Oct., 1926), 97-106.

222. Escaille, Capitaine de Vaisseau de l'. "L'aéronautique maritime." R.F.A. 8 (Mar., 1930), 301-319.

223. ________. "Pour une politique industrielle de l'aéronautique." R.A.A. 76 (Nov., 1935), 1203-1210.

224. ________. "Deux points de vue sur la politique industrielle en aéronautique." R.A.A. 84 (July, 1936), 727-734.

225. "L'esprit chevaleresque des aviateurs." L.A. 586 (Sept. 8, 1932), 6; 587 (Sept. 15, 1932), 7; 588 (Sept. 22, 1932), 11; 589 (Sept. 29, 1932), 7; 590 (Oct. 6, 1932), 6; 591 (Oct. 13, 1932), 7.

226. Estailleur-Chanteraine, Philippe d'. "Il y a un quart de siècle, le premier tour aérien d'Afrique." F.A.F. 115 (May, 1936), 863-884.

227. Estienne, Lieutenant. "Comment réaliser un service aérien à travers le Sahara." R.A.M. 11 (Sept.-Oct., 1922), 97-100.

228. Etienne, Pierre. "Le problème de l'avion de reconnaissance stratégique." R.F.A. 22 (May, 1931), 513-544.

229. ________. "Aviation et unités motorisées." R.A.A. 75 (Oct., 1935), 1083-1104.

230. ________. "Réserve générale et division aérienne." R.A.A. 76 (Nov., 1935), 1257-1269.

231. ________. "Contribution à l'étude de la guerre aérienne autonome." R.A.A. 77 (Dec., 1935), 1323-1336.

232. ________. "La doctrine officielle de l'aviation militaire française." R.A.A. 79 (Feb., 1936), 123-129.

233. ________. "Esquisse d'une doctrine de l'avion projectile." R.A.A. 80 (Mar., 1936), 243-251.

234. ________. "L'attaque aérienne d'un grand centre." R.A.A. 99 (Oct., 1937), 1135-1140.

235. ________. "Secteurs de feux." R.A.A. 109 (Aug., 1938), 858-870.

236. ________. "Transports aériens militaires." R.A.A. 116 (May-June, 1939), 263-268.

237. Etévé, Inspecteur Général Albert. "Les avions et les moteurs en 1918." F.A.F. 252 (Nov., 1968), 491-512.

238. "L'étude des problèmes d'aviation d'arrêt." R.A.A. 92 (Mar., 1937), 243-248.

239. Eynac-Laurent. "Le réarmement aérien français." REVUE POLITIQUE ET PARLEMENTAIRE, Oct. 10, 1938, p. 16.

240. "L'expérience actuelle. Les terrains de travail de l'aviation au Maroc pendant les opérations 1925-1926. (Extraits du rapport du colonel Armengaud)." R.A.M. 42 (Nov.-Dec., 1927), 132.139.

241. Facon, Patrick. "L'aviation dans une revue interarmes de l'entre deux guerres: 1923-1938." R.H.A. (No. 2, 1977), 103-112.

242. ________. "Quand la France achetait ses avions aux USA: 1938-1940." AVIATION MAGAZINE INTERNATIONAL 757 (July 1, 1979), 87-91; 758 (July 15, 1979), 83-87.

243. ________. "Aux origines du réarmement aérien français: le plan I, 1934-1937." AVIATION MAGAZINE INTERNATIONAL 747 (Feb. 1, 1979), 86-91; 748 (Feb. 15, 1979), 66-71.

244. ________. "L'aviation populaire: entre les mythes et la realité." R.H.A. (No. 2, 1982), 54-66.

245. ________. "La visite du Général Vuillemin en Allemagne (16-21 aout 1938)." R.H.A. (No. 2, 1982), 110-121.

246. ________. "Le haut commandement aérien français et la crise de Munich." R.H.A. (No. 3, 1983), 10-17.

247. Faure, Pierre. VERS UN NOUVEAU CHARLEROI. LA GUERRE AERIENNE DE DEMAIN. Paris: 1931.

248. ________. L'ARMEE DE L'AIR ET LA SECURITE DE LA FRANCE. Paris: 1932.

249. ________. L'AVION TUERA LA GUERRE... Paris: 1935.

250. Feldzer, V. "Etude sur la défense anti-aérienne." R.A.A. 98 (Sept., 1937), 1001-1018.

251. Ferrus, Lieutenant-Colonel. "L'aviation coloniale." L'AEROPHILE 17-18 (Sept. 1-15, 1919), 265-267.

252. Fischetti, Ugo. "Dottrina ed organizaione dell'aeronautica di Francia." RIVISTA AERONAUTICA, 8 (Mar., 1932), 399-409.

253. Fichot, Commandant E. "Quelques réflexions sur l'observation aérienne du champ de bataille." R.A.A. 106 (May, 1938), 526-546.

254. Flamme, Médecin Commandant. "Les fatigues exceptionnelles du service aérien." R.A.A. 62 (Sept., 1934), 987-1029.

255. Flandin, Pierre Etienne. "Une politique aérienne." JOURNAL OFFICIEL, June 4, 1926.

256. Fleury, Colonel Jean. "Dix ans après." DEFENSE NATIONALE, June, 1979.

257. Fonck, René. L'AVIATION ET LA SECURITE FRANCAISE. Paris: 1924.

258. Fontaine, G. Chef d'escadron. "L'artillerie anti-aérienne." R.A.M. 3 (May-June, 1921), 59-68; 5 (Sept.-Oct., 1921), 100-106.

259. ________. "Historique sommaire de la défense contre aéronefs." R.A.M. 1 (Jan-Feb., 1922), 14-20; 2 (Mar.-Apr., 1922), 38-40.

260. Franck, Capitaine P. "Les liaisons par T.S.F. dans l'aviation militaire." R.A.M. 8 (Mar.-Apr., 1922), 36-41.

261. Frankenstein, Robert. "A propos des aspects financiers du réarmement française (1935-1939)." REVUE D'HISTOIRE DE LA DEUXIEME GUERRE MONDIALE 102 (April, 1976), 1-20.

262. ________. "Le financement du réarmement français (1935-1939)." Unpublished thesis, doctorat de troisième cycle, Université de Paris I, 1980.

263. Fridenson, Patrick. "Forces et faiblesses des conversations aériennes franco-britanniques." LES RELATIONS FRANCO-BRITANNIQUES DE 1936 A 1939. Paris: 1975, 359-373.

264. Fridenson, Patrick and Jean Lecuir. "L'aviation dans les projets franco-britanniques, 1935-1938." LES RELATIONS FRANCO-BRITANNIQUES DE 1936 A 1939. Paris: 1975, 119-126.

265. ________. LA FRANCE ET LA GRANDE-BRETAGNE FACE AUX PROBLEMES AERIENS (1935-MAI 1940). Vincennes: 1976.

266. Gallois, Général P.M. "1939. Doctrine française de guerre aérienne." COLLOQUE INTERNATIONAL SUR L'ADAPTATION DE L'ARME AERIENNE AUX CONFLITS CONTEMPORAINS ET PROCESSUS D'INDEPENDANCE DES ARMEES DE L'AIR. Paris: 1985, 47-63.

267. Garnier, R. "Problèmes de la supériorité balistique du tir retraite." R.A.A. 86 (Sept., 1936), 971-976.

268. Garsonnin, Capitaine. "Le monoplace, avion de renseignement et de travail." R.A.A. 65 (Dec., 1934), 1347-1377.

269. ________. "Le monoplace comme appareil de coopération." R.A.A. 106 (May, 1938), 547-574.

270. Gaudillère, Commandant. "Defense de l'Empire et escadrilles impériales." R.A.A. 116 (May-June, 1939), 243-262.

271. Gaulmier, P. Lieutenant. "Chasse de nuit. Monoplace ou Biplace?" R.A.A. 70 (May, 1935), 524-534.

272. Gérardot, Général P. "L'armement aérien de la France entre 1934 et 1940: Quelles furent les causes de l'échec total du Plan 1?" L.A. 1307 (Feb., 1951), 6.

273. ________. "L'armement aérien de la France entre 1934 et 1940: Ce qu'apportèrent à notre aviation le plan quinquenal et le Plan II." L.A. 1308 (Feb. 24, 1951), 6.

274. ________. "L'armement aérien de la France entre 1934 et 1940: Le plan V ou plan de 2600 fut entrepris sous le signe de la lenteur et des hésitations." L.A. 1309 (Mar. 3, 1951), 6.

275. ________. "L'armement aérien de la France entre 1934 et 1940: L'état-major de l'Air s'aperçut, un peu tard qu'il ne peuvent compter que sur ses propres moyens." L.A. 1310 (Mar. 10, 1951), 6.

276. ________. "L'armement aérien de la France entre 1934 et 1940. La crise du personnel fut aussi grave que celle qui frappa le matériel." L.A. 134 (Mar. 17, 1951), 6.

278. Goussot, Lieutenant C.M. "L'organisation des sections de photographie aérienne." R.A.M. 16 (July-Aug., 1923), 74-76; 18 (Nov.-Dec., 1923), 136-138; 21 (May-June, 1924), 67-70.

279. ________. "L'organisation du travail dans une section de photographie aérienne en temps de paix." R.A.M. 25 (Jan.-Feb., 1925), 14-16.

280. "Un grand débat parlementaire sur la défense nationale." R.A.A. 93 (Apr., 1937), 423-451.

281. "Les grands raids et les grands voyages effectués par l'aviation militaire en 1926." R.A.M. 36 (Nov.-Dec., 1926), 136-140.

282. "Les grands voyages aériens; France-Madagascar." R.A.M. 38 (Mar.-Apr., 1927), 30-33.

283. Gravellat, Médecin Major. "L'aviation sanitaire aux colonies." R.A.M. 18 (Nov.-Dec., 1923), 122-124.

284. Grigaut, Général. "L'école de l'air." R.H.A. Special aviation issue, 1969, 151-168.

285. Gruel, Commandant. "L'emploi de la TSF dans l'aéronautique." R.A.M. 37 (Jan.-Feb., 1927), 6-11.

286. Gruss, Robert. LES FLOTTES DE L'AIR EN 1937. Preface by Colonel A. Pinsard. Paris: 1937.

287. _________. LES FLOTTES DE L'AIR EN 1938. Paris: 1938.

288. "La guerre aérienne au cinéma." L.A. 388 (Nov. 22, 1928), 11.

289. "La guerre aérienne en Espagne." R.A.A. 91 (Feb., 1937), 185-198.

290. "La guerre aérienne en Espagne." R.A.A. 96 (July, 1937), 808-824.

291. LA GUERRE AERIENNE. REVISION DES LOIS DE LA GUERRE. The Hague: 1930.

292. "La guerre chimique. Rapport officiel in la Société des Nations établi par la Commission temporaire mixte pour la réduction des armements." L.A. 195 (Feb. 12, 1925), 3-4.

293. Guet, Capitaine. "Les causes d'accidents graves dans les entreprises de transports aériens subventionnées et dans les forces aériennes de terre." R.F.A. 28 (Nov., 1931), 1337-1351; 29 (Dec., 1931), 1454.

294. Guillemeney, Colonel. "L'aviation est-elle apte à la guerre de mouvement?" R.F.A. 2 (Sept., 1929), 195-223.

295. ________. "Vieux principes. Méthodes nouvelles." R.F.A. 27 (Oct., 1931), 1189-1216.

296. Gunsburg, Jeffery, A. DIVIDED AND CONQUERED: THE FRENCH HIGH COMMAND AND THE DEFEAT OF THE WEST, 1940. Westport, Conn.: 1979

297. Guyomar, Lieutenant-Colonel. "L'aviation de bombardement moyens porteurs dans la bataille de jour." R.F.A. 31 (Feb., 1932), 155-174.

298. Haight, John McVickar. AMERICAN AID TO FRANCE, 1938-1940. New York: 1970.

299. ________. "France's First War Mission to the U.S." A.H. (Summer, 1964), 11-15.

300. ________. "France's Search for American Military Aircraft: Before the Munich Crisis." A.H. (Fall, 1978), 141-152.

301. ________. "Les négociations françaises pour la fourniture d'avions américains avant Munich." F.A.F. 198 (Dec., 1963), 807-839.

302. ________. "Les négociations relatives aux achats d'avions américains par la France pendant la période qui précède immédiatement la guerre." REVUE D'HISTOIRE DE LA DEUXIEME GUERRE MONDIALE 48 (Apr., 1965), 1-34.

303. Hébrard, Capitaine. "De l'aviation de bombardement de nuit." R.A.M. 22 (July-Aug., 1924), 87-90.

304. Hébrard, Lieutenant-Colonel. "L'aviation de nuit dans le passé, le présent et l'avenir." R.F.A. 42 (Jan., 1933), 5-60; 43 (Feb., 1933), 123-150; 44 (Mar., 1933), 245-275.

305. Henot, Capitaine P. "La chasse et le futur combat aérien." R.F.A. 36 (July, 1932), 771-778.

306. Henry Couannier, André. LEGITIMITE DE LA GUERRE

AERIENNE. OPINIONS RECUEILLIS. Paris: 1925.

307. Herbillon, Jean. "Avion d'abord, béton ensuite." L.A. 398 (Jan. 31, 1929), 5.

308. ________. "L'offensive de M. Guy de Lubersac. Le sort du Ministère de l'Air." L.A. 402 (Feb. 28, 1929), 5.

309. ________. "La vie des idées. Les 'Turcs' dans l'Armée de l'Air." L.A. 412 (May 9, 1929), 5.

310. ________. "En marge des déclarations au ministère de l'Air." L.A. 415 (May 30, 1929), 5.

311. ________. "Le statut du personnel: Pensons à nos sous-officiers." L.A. 41 (June 27, 1929), 5.

312. ________. "La défense de l'intégral. Bon livre et bonne action." L.A. 421 (July 11, 1924), 5.

313. ________. "La tenue, l'esprit de l'infanterie et autres." L.A. 429 (Sept. 5, 1929), 5.

314. ________. "Premier regard sur le 'problème aérien'." L.A. 431 (Sept. 19, 1929), 5.

315. ________. "La recherche du temps perdu." L.A. 437 (Oct. 31, 1929), 5.

316. ________. "La mode...aviateurs type '29'." L.A. 438 (Nov. 7, 1929), 5.

317. ________. "Première analyse du statut du personnel." L.A. 450 (Jan. 30, 1930), 5.

318. ________. "Les lois organiques. M. de La Palisse attaque le malin." L.A. 151 (Feb. 6, 1930), 5.

319. ________. "Les grands débats. Les forces aériennes à la Chambre des Députés." L.A. 452 (Feb. 13, 1930), 5.

320. ________. "Le problème des réserves est posé." L.A. 453 (Feb. 20, 1930), 5.

321. ________. "L'organisation générale des forces

aériennes." L.A. 456 (Mar. 13, 1930), 5.

322. ________. "Théorie générale de la destruction." L.A. 461 (Apr. 17, 1930), 5.

323. ________. "Le statut des sous-officiers." L.A. 466 (May 25, 1930), 5.

324. ________. "La formation du personnel navigant." L.A. 471 (June 26, 1930), 5.

325. ________. "Le danger aérien et le bon français." L.A. 475 (July 24, 1930), 5.

326. ________. "Le bon français aborde le général Douhet." L.A. 477 (Aug. 7, 1930), 5.

327. ________. "Les préliminaires de la guerre de 19.." L.A. 479 (Aug. 21, 1930), 5.

328. ________. "La journée du 16 juin 19.." L.A. 480 (Aug. 28, 1930), 5; 481 (Sept. 9, 1930), 5; 482 (Sept. 11, 1930), 5; 483 (Sept. 18, 1930), 5; 485 (Oct. 2, 1930), 5; 487 (Oct. 16, 1930), 5.

329. ________. "L'oeuvre remarquable du colonel Vauthier." L.A. 484 (Sept. 25, 1930), 7.

330. ________. "Le bon français saura-t-il oser?" L.A. 488 (Oct. 28, 1930), 5.

331. ________. "Une politique de l'usine." L.A. 489 (Oct. 30, 1930), 5.

332. ________. "Les leçons de l'histoire." L.A. 495 (Dec. 11, 1930), 13.

333. ________. "Le 'Crime' du ministère de l'Air." L.A. 498 (Jan. 1, 1931), 5.

334. ________. "Sur l'unité de commandement." L.A. 503 (Feb. 5, 1931), 5.

335. Herre, Paul. "Le bombardier et la chasseur de nuit." L.A. 417 (June 13, 1929), 6.

336. Hirschauer, Général. "Les Ailes au Sénat. Le

développement de l'Aéronautique suit le progrès technique." L.A. 106 (June 28, 1923), 4.

337. Hodeir, Marcellin. "La chambre bleu horizon et la commission de l'aviation militaire (1919-1924)." COLLOQUE INTERNATIONAL SUR L'ADAPTATION DE L'ARMEE AERIENNE AUX CONFLITS CONTEMPORAINS ET PROCESSUS D'INDEPENDANCE DES ARMEES DE L'AIR. Paris: 1985, 341-352.

338. "Les hostilités en cours et les operations de guerre." R.A.A. 119 (Nov.-Dec., 1939), 607-626.

339. Houard, Georges. "Ce qu'ils font au Maroc." L.A. 25 (Dec. 8, 1921), 1.

340. _________. "Un ministre, deux présidents." (L.A. 31 (Jan. 19, 1922), 1.

341. _________. "Il faut développer l'aviation coloniale." L.A. 31 (Jan. 19, 1922), 1.

342. _________. "L'aéronautique n'occupe pas une place suffisante dans notre marine." L.A. 51 (June 8, 1922), 1.

343. _________. "Le colonel Saconney. Autour d'une démission." L.A. 59 (Aug. 3, 1922), 1.

344. _________. "Il faut étudier le parachute." L.A. 105 (June 10, 1923), 1.

345. _________. "Armements aériens." L.A. 108 (July 10, 1923), 1.

346. _________. "A la veille d'une crise." L.A. 127 (Nov. 22, 1923), 1, 4.

347. _________. "L'aviation ne 'paie pas'." L.A. 128 (Nov. 29, 1923), 1.

348. _________. "Mais l'aviation ne pourrait-elle pas 'payer'?" L.A. 129 (Dec. 6, 1923), 1.

349. _________. "Avions de bombardement, avions de transport." L.A. 142 (Mar. 6, 1924), 1.

350. ________. "La dernière alerte." L.A. 146 (Apr. 3, 1924), 1.

351. ________. "Aurons-nous notre ministère?" L.A. 153 (May 22, 1924), 1.

352. ________. "La liberté de construire." L.A. 160 (July 10, 1924), 1.

353. ________. "Désarmement." L.A. 168 (Sept. 4, 1924), 1.

354. ________. "On peut réduire le risque aérien." L.A. 174 (Oct. 16, 1924), 1.

355. ________. "Les Pilotes de l'Armée." L.A. 179 (Oct. 23, 1924), 1.

356. ________. "Le corps des ingénieurs." L.A. 179 (Nov. 20, 1924), 1.

357. ________. "A coté du but..." L.A. 180 (Nov. 27, 1924), 1.

358. ________. "La menace aéro-chimique." L.A. (Jan. 22, 1925), 1.

359. ________. "Politique aérienne franco-anglaise." L.A. 194 (Mar. 5, 1925), 1.

360. ________. "La leçon du Maroc." L.A. 205 (May 21, 1925), 1.

361. ________. "Que la leçon nous serve!" L.A. 210 (June 25, 1925), 1.

362. ________. "La continuité dans l'effort." L.A. 221 (Sept. 16, 1925), 1.

363. ________. "Une arme? Qui! au moins." L.A. 264 (July 8, 1926), 1.

364. ________. "Ceci remplacera cela." L.A. 265 (July 16, 1926), 1.

365. ________. "La faute." L.A. 267 (July 29, 1926), 1.

366. ________. "Vers un jeu de dupes..." L.A 273 (Sept. 9,

1926), 1.

367. ________. "L'école de l'aviation." L.A. 278 (Oct. 14, 1926), 1.

368. ________. "La nouvelle marine." L.A. 284 (Nov. 27, 1926), 1.

369. ________. "Les gaz et l'avion." L.A. 297 (Feb. 24, 1927), 1.

370. ________. "L'aviation dans la guerre." L.A. 336 (Nov. 24, 1927), 1.

371. ________. "Le danger aérien est-il réel? L'aviation dans la guerre." L.A. 340 (Dec. 22, 1927), 3.

372. ________. "Sous les fleurs." L.A. 343 (Jan. 12, 1928), 1.

373. ________. "Un homme, un programme." L.A. 344 (Jan. 19, 1928), 1.

374. ________. "La lutte pour le ministère de l'Air." L.A. 380 (Sept. 27, 1928), 1, 2.

375. ________. "Le monopole de la conception est un frein à l'émulation technique." L.A. 385 (Nov. 1, 1928), 1, 2.

376. ________. "L'armée de l'Air: problème national." L.A. 403 (Mar. 7, 1929), 1, 2.

377. ________. "L'effort technique et industriel du Ministère de l'Air." L.A. 434 (Oct. 10, 1929), 7, 8.

378. ________. "Quelques points essentiels d'une politique de l'Air." L.A. 437 (Oct. 31, 1929), 1, 2, 4.

379. ________. "L'aviation, la paix et la guerre." L.A. 483 (Sept. 18, 1930), 1, 2.

380. ________. "La politique aérienne de la France. Les Ailes ont interviewé M. Etienne Riché." L.A. 525 (July 9, 1931), 4.

381. ________. "Supprimez la guerre, ne supprimez pas l'aviation." L.A. 556 (Feb. 11, 1932), 1, 2.

382. ________. "Appel au Parlement. La suppression du Ministère de l'Air." L.A. 558 (Feb. 25, 1932), 1, 2.

383. ________. "En pleine crise. La désorganisation de l'Air." L.A. 559 (Mar. 3, 1932), 1-3.

384. ________. "La cité de l'Air est édifiée." L.A. 656 (Jan. 11, 1934), 9, 10.

385. ________. "Notre matériel militaire est rénové. Mais comment maintenir sa qualité." L.A. 741 (Aug. 29, 1935), 8-9.

386. ________. "La 'nationalisation' de l'industrie aéronautique met en péril le développement de notre aviation." L.A. 785 (July 2, 1956), 13.

387. ________. "L'étatisation totale et prochaine des usines d'aviation livrera toute la technique aéronautique aux services officiels." L.A. 786 (July 9, 1936), 13.

388. ________. "Comment M. Pierre Cot envisage la 'nationalisation'." L.A. 787 (July 17, 1936), 12.

389. ________. "La Chambre a voté la 'nationalisation'." L.A. 788 (July 23, 1936), 13.

390. ________. "Comment le ministère de l'Air conçoit l'aviation populaire." L.A. 791 (Aug. 13, 1936), 12.

391. ________. "Nous avons un nouveau Ministre de l'Air." L.A. 867 (Jan. 27, 1938), 3.

392. ________. "Six avions par jour? La France peut produire plus!" L.A. 882 (May 12, 1938), 3.

393. ________. "Renforçons d'abord notre aviation mais, en même temps pensons à faire la paix." L.A. 904 (Oct. 13, 1938), 3.

394. ________. "Fallait-il ou non acheter les P.36?" L.A.

920 (Feb. 2, 1939), 3.

395. ________. "Albert Caquot, 'dictateur à la production'." L.A. 951 (Oct. 29, 1939), 5.

396. ________. "Le général Denain." L.A. 1406 (June 10, 1953), 5.

397. Ide, John Jay. "The French Air Ministry." AIRWAY AGE 12 (No. 2, 1931), 149-151; 12 (No. 3, 1931), 238-241.

398. ________. "Two Years of the French Air Ministry." NINETEENTH CENTURY, 109 (Feb., 1931), 148-161.

399. Izard, L., J. Des Cilleuls, and R. Kermarrec. LA GUERRE AERO-CHIMIQUE ET LES POPULATIONS CIVILES: ETUDE HISTORIQUE, CHIMIQUE, THERAPEUTIQUE ET PREVENTIVE. Preface by General Niessel. Paris: 1932.

400. Jauneaux, Commandant Marcel. L'EVOLUTION DE L'AERONAUTIQUE. Paris: 1923.

401. ________. L'AVIATION MILITAIRE ET LA GUERRE AERIENNE. Paris: 1923.

402. Joscat, René. SIRENE D'ESCADRILLE. ROMAN D'UNE ESCADRILLE. Foreword by Achille Segard. Paris: 1932.

403. Jousseaume, Commandant. "L'emploi nocturne du ballon d'observation." R.A.A. 96 (July, 1937), 799-805.

404. Julliot, CH. L. "Les avions sanitaires." L.A. 12 (Sept. 8, 1921), 1; 19 (Sept. 27, 1921), 1; 27 (Dec. 22, 1921), 1; 32 (Jan. 26, 1922), 1; 47 (May 11, 1922), 1; 51 (June 8, 1922), 3; 56 (July 13, 1922), 4; 60 (Aug. 10, 1922), 3; 65 (Sept. 14, 1922), 1, 4; 68 (Oct. 5, 1922), 3.

405. ________. "L'aviation sanitaire au Maroc." L.A. 208 (June 11, 1925), 4; 209 (June 18, 1925), 4.

406. Kane, Francis X. "French Airpower in the Thirties." A.H. (Autumn, 1966), 105-113.

407. Keller, Chef de bataillon. "L'aviation dans la bataille." R.A.M. 27 (May-June, 1925), 50-55.

408. Kessel, J. L'EQUIPAGE. Paris: 1923.

409. ________. LE REPOS DE L'EQUIPAGE. Paris: 1935.

410. Kitcheeff, Georges. "Conditions d'efficacité stratégique de l'aviation." R.A.A. 109 (Aug., 1938), 843-857.

411. ________. "Le problème de la guerre en Europe." R.A.A. 112 (Nov., 1938), 1203-1222.

412. ________. L'AVIATION POURRA-T-ELLE NOUS EVITER LA GUERRE? Dunkirk: 1938.

413. Kohl, Wilfred L. "The French Nuclear Force and Alliance Diplomacy, 1958-1967." Unpublished dissertation, Columbia University, 1968.

414. Krauskopf, Robert W. "French Air Power Policy, 1919-1939." Unpublished dissertation, Georgetown University, 1965.

415. L.A. "L'aviation militaire colonial." L.A. 249 (Mar. 25, 1926), 3.

416. Labaurie, Capitaine. "Un équipage français autour de la baltique." R.A.M. 42 (Nov.-Dec., 1927), 128-131.

417. Labric, Roger. LES CHAMPS BLEUS. Paris: 1923.

418. ________. ON SE BAT DANS L'AIR. Preface by R. Fonck. Paris: 1936.

419. Lafon, Commandant Charles. ETUDE SUR LES BALLONS CAPTIFS ET LES AERONEFS MARINS. APPLICATIONS AUX NAVIRES DE SURFACE. Paris: 1922.

420. La Grange, Sénateur A de. "L'Industrie aéronautique et la défense nationale." R.A.A. 93 (Apr., 1937), 365-382.

421. Lainé, Serge. "L'Aéronautique militaire française au

Maroc (1911-1939)." R.H.A. (No. 4, 1978), 107-119.

422. Lallier, Roger. EN AVION AU MAROC. HOMMAGE AU PILOTE. Paris: 1923.

423. Lamarche, Paul E. "French Pursuit Planes." AERO DIGEST 17, No. 1 (July, 1930), 128-131ñ 17, No. 2 (Aug., 1930), 92-95.

424. Lamarzelle, Capitaine de. "L'école de l'Air." R.A.A. 119 (Nov.-Dec., 1939), 633-639.

425. Lamy, Paul. "Pelletier d'Oisy." L.A. 57 (July 20, 1922), 1.

426. Laporte, J.F. "Faut-il imposer le parachute? Le parachute est indispensable aux avions militaires." L.A. 133 (Jan. 3, 1924), 1, 3.

427. Langeron, André. "Istres, école des sous-officiers de l'Air." L.A. 624 (June 1, 1933), 8.

428. ________. "Comment est organisée la belle école d'Istres." L.A. 625 (June 8, 1933), 10.

429. ________. "Comment sont formés pilotes et mitrailleurs." L.A. 627 (June 22, 1933), 8.

430. ________. "A Rochefort avec les apprentis mécaniciens de l'Air." L.A. 632 (July 27, 1933), 8.

431. ________. "L'aviation nouvelle. L'oeil du maitre." L.A. 576 (June 30, 1932), 5.

432. ________. "Le désarmement aérien. Les ombres inquiètantes." L.A. 584 (Aug. 25, 1932), 5, 6.

433. ________. "Prôblèmes budgétaires." L.A. 585 (Sept. 1, 1932), 5.

434. ________. "L'aviation et la Parlement." L.A. 589 (Sept. 29, 1932), 4.

435. ________. "L'aviation et l'opinion." L.A. 591 (Oct. 13, 1932), 4.

436. ________. "La malaise militaire." L.A. 665 (Mar. 15, 1934), 7.

437. ________. "Il faut créer un 'Douhet français'." L.A. 677 (June 7, 1934), 11.

438. ________. "Les préoccupations du Parlement sur la politique aérienne de la France." L.A. 681 (July 5, 1934), 11.

439. ________. "L'organisation de la défense nationale et la politique industrielle de l'Air." L.A. 682 (July 12, 1934), 15, 16.

440. ________. "Les futurs programmes de matériels militaires et la liberté technique." L.A. 685 (Aug. 2, 1934), 11.

441. ________. "Considérations sur l'aviation d'arrêt." L.A. 707 (Jan. 3, 1935), 7.

442. ________. "L'opinion du Président de la Commission de l'Aéronautique sur le statut du personnel de l'Armée de l'Air." L.A. 709 (Jan. 17, 1935), 11.

443. ________. "Le statut du personnel. I-Principes et raisons d'être du statut." L.A. 711 (Jan. 31, 1935), 11-12; "II-grandeur et misères des sous-officiers de l'air," 713 (Feb. 14, 1935), 13-14; "III-Remarques sur le statut des officiers navigants," 715 (Feb. 25, 1935), 13-14; "IV-Le recrutement des officiers de l'air," 718 (Mar. 21, 1935), 3; "V-L'avancement des officiers navigants," 720 (Apr. 4, 1935), 6-7.

444. ________ and Jean Romeyer. "Coup d'oeil sur l'aviation d'assaut du Colonel Mecozzi." L.A. 712 (Feb. 7, 1935), 11-12.

445. ________. "La volonté de surprise. L'armée de l'Air à la croisée des chemins." L.A. 717 (Mar. 14, 1935), 7.

446. ________. "La pensée française devant Douhet." L.A. 719 (Mar. 28, 1935), 8.

447. ________. "L'organisation aérienne de la France."

L.A. 722 (Apr. 18, 1935), 8.

448. ________. "Le plan des dix mille avions-1." L.A. 725 (May 9, 1935), 6, 7; "2-Appel à la compréhension des dirigeants," 728 (May 30, 1935), 7; "3-Appel à la compréhension de l'opinion," 730 (June 13, 1935), 8; "4-Plus d'efficacité dans l'action pour moins d'argent," 731 (June 20, 1935), 12; "5-Pour une politique de preuves," 736 (July 25, 1935), 6; "6-La multiplication du prototype," 738 (Aug. 8, 1935), 6; "7-L'attaque des forces vives," 740 (Aug. 22, 1935), 7; "8-La formation du personnel navigant exige des solutions nouvelles," 741 (Aug. 29, 1935), 7.

449. ________. "La politique aérienne de la France." L.A. 743 (Sept. 12, 1935), 6.

450. ________. "L'organisation de la désorganisation. Quelques questions sur la 'lutte aérienne' et la 'coopération'." L.A. 744 (Sept. 19, 1935), 6.

451. ________. "Les '80' pourraient-ils gagner la guerre? Le divorce de la conception et de l'exécution." L.A. 763 (Jan. 30, 1936), 6.

452. ________. "C'est l'inscription aérienne qui nous donnera le corps des navigants." L.A. (Feb. 6, 1936), 6, 7.

453. ________. "Qui sera le chef suprème en 1940?" L.A. 765 (Feb. 13, 1936), 7.

454. ________. "Dans l'Europe inquiète, l'aviation va jouer le premier rôle. Aux sources d'une science nouvelle: l'aérostratégie." L.A. 771 (Mar. 26, 1936), 6.

455. ________. "Vers un ministère de la defense nationale." L.A. 778 (May 19, 1936), 6.

456. ________. "La vérité sur le plan quinquennal." L.A 793 (Aug. 27, 1936), 7.

457. ________. "L'équation de notre puissance aérienne. 1-Les trois insuffisances du plan des '1010'," L.A. 794 (Sept. 3, 1936), 7; "2-Il faudrait

10.000 avions; on en prévoit 1500," 795 (Sept. 10, 1936), 7.

458. ________. "L'armée de l'Air serait notre seule riposte efficace. Or elle est notoirement insuffisante." L.A. 748 (Oct. 1, 1936), 7.

459. ________. "L'échec du multiplace de la doctrine '1933'." L.A. 800 (Oct. 15, 1936), 7.

460. ________. "La nouvelle mission: arrêter l'invasion brusquée." L.A. 807 (Dec. 3, 1936), 7.

461. ________. "Le Ministère de la défense nationale doit devenir sans retard une réalité." L.A. 808 (Dec. 10, 1936), 7.

462. ________. "L'organisation aérienne de la France doit être étayée et controlée." L.A. 809 (Dec. 17, 1936), 7.

463. ________. "La valeur des cadres navigants est notre suprème espoir." L.A. 814 (Jan. 21, 1937), 12.

464. ________. "Peu d'étoffe, trop de cumuls dans le Haut Commandement." L.A. 816 (Feb. 4, 1937), 13.

465. ________. "L'armée de l'Air vue par son chef." L.A. 817 (Feb. 11, 1937), 12.

466. ________. "Le Potez 63 est-il un avion d'arrêt?" L.A. 822 (Mar. 18, 1937), 13.

467. ________. "Que doit être l'avion d'arrêt idéal?" L.A. 824 (Mar. 1, 1937), 12.

468. ________. "La guerre d'Espagne et ses leçons." L.A. 825 (Apr. 8, 1937), 12.

469. ________. "La vraie coopération peut s'établir entre l'armée de terre et l'aviation." L.A. 826 (Apr. 15, 1936), 13.

470. ________. "L'attaque d'Avord. Au quatorzième jour seulement l'infanterie succomba." L.A. 826 (Mar. 4, 1937), 13.

471. ________. "Controverse sur la doctrine. Attaquer l'ennemi partout où il se trouve." L.A. 828 (Apr. 29, 1937), 13.

472. ________. "La légion aérienne parachutée, armée redoutable et redoutée." L.A. 829 (May 6, 1937), 12.

473. ________. "L'Air intégral, nouvelle formule." L.A. 832 (May 27, 1937), 12.

474. ________. "La coopèration de l'aviation et des autres en armes doit être envisagée dans un cadre nouveau." L.A. 834 (June 10, 1937), 13.

475. ________. "Toujours plus haut! I-Première esquisse d'une doctrine de l'altitude," L.A. 835 (June 17, 1937), 11; "II-Comment pourra-t-on opérer entre 6000 et 10.000 mètres?" 837 (July 1, 1937), 11; "III-Reconnaissance d'un nouvel ennemi: les 'G'," 839 (July 15, 1937), 11; "IV-Les effets des accélérations," 844 (Aug. 19, 1937), 11; "V-Préoccupations du médecin d'aviation." L.A. 845 (Aug. 26, 1937), 11.

476. ________. "L'échec du multiplace de la doctrine '1933'." L.A. 800 (Oct. 15, 1936), 7.

477. ________. "La nouvelle mission: arrêter l'invasion brusquée." L.A. 807 (Dec. 3, 1936), 7.

478. ________. "L'organisation de notre défense aérienne." L.A. 836 (June 24, 1937), 12.

479. ________. "A propos d'un jugement sur le 'plan des 10.000'." L.A. 843 (Aug. 12, 1937), 7.

480. ________. "Les besoins de l'Armée de l'Air imposent un gros effort industriel." L.A. 846 (Sept. 2, 1937), 11.

481. ________. "Critique des dernières manoeuvres." L.A. 847 (Sept. 9, 1937), 11.

482. ________. "Grandeurs et difficultés de l'Armée de l'Air, I." L.A. 849 (Sept. 23, 1937), 11; "II-Cinq personnalités éminentes, cinq options

différentes," 851 (Oct. 7, 1937), 11; "III-La faute des corps techniques," 852 (Oct. 14, 1937), 11; "IV-L'ingénieur d'état se défend," 856 (Nov. 11, 1937), 11; "V-Le tir, acte essentiel, a été oublié!" 858 (Nov. 25, 1937), 11.

483. ________. "Le général Denain, grand chef de l'Air." L.A. 850 (Sept. 30, 1937), 9.

484. ________. "L'armée de l'Air devant le Parlement. On ne fait pas la guerre avec des espérances." L.A. 862 (Dec. 23, 1937), 11.

485. ________. "Verra-t-on de nouveau se poser la question du Ministère de l'Air." L.A. 863 (Dec. 30, 1937), 11.

486. ________. "Une interview de notre premier ministre de l'Air. Dans la défense nationale, c'est sur l'Air qu'il faut mettre l'accent nous dit M. Laurent-Eynac." L.A. 870 (Feb. 17, 1938), 6.

487. ________. "Cinq auteurs à la recherche d'une politique de l'Air." L.A. 873 (Mar. 10, 1938), 11.

488. ________. "Dans le ciel d'Europe, l'aviation grandit démesurément." L.A. 884 (May 26, 1938), 11.

489. ________. "De la commande des cent Curtiss à la mission du général Vuillemin." L.A. 886 (June 9, 1938), 11.

490. ________. "Tout pour la chasse! Parfait...mais que vaut la chasse?" L.A. 887 (June 16, 1938), 11.

491. ________. "La recontre en Allemagne des 'chevaliers sans haine'." L.A. 898 (Sept. 1, 1938), 5, 7.

492. ________. "L'organisation de l'armée de l'Air. Le retour à la position de juin 1936." L.A. 900 (Sept. 15, 1938), 11.

493. ________. "Réveillons les énergies aériennes." L.A. 904 (Oct. 13, 1938), 11.

494. ________. "Ce sont les besoins des forces armées qui

doivent fixer l'importance de notre aviation." L.A. 907 (Nov. 3, 1938), 11.

495. ________. "Il faut que nous construisions les avions que notre situation impose." L.A. 908 (Nov. 10, 1938), 11.

496. ________. "Le bilan de l'année 1938 montre qu'il y aura beaucoup à faire en 1939." L.A. 916 (Jan. 5, 1939), 11.

497. ________. "C'est dans l'usine, l'escadre, l'école que réside la vraie puissance aérienne." L.A. 921 (Feb. 9, 1939), 11.

498. ________. "L'inscription aérienne de la jeunesse française." L.A. 923 (Feb. 23, 1939), 11.

499. ________. "La stratégie technique, base de l'équilibre des forces aériennes sur nos frontières." L.A. 926 (Apr. 6, 1939), 11.

500. ________. "Faire face. 1-Laisserons nous les allemands maitres du ciel en Europe?" 934 (May 11, 1939), 11; "2-La force franco-anglaise doit retrouver la supériorité dans le domaine de l'air," 935 (May 18, 1939), 11; "3-La chasse est une arme difficile qui demande des pilotes exceptionnels," 938 (June 8, 1939), 11; "4-Un rude effort a été fait mais il n'est pas encore suffisant," 941 (June 29, 1939), 11; "5-Il faut dire 'oui' aux chercheurs," 942 (July 6, 1939), 11; "6-Il ne doit pas exister de dualité entre les utilisateurs et les techniciens," 943 (July 13, 1939), 11.

501. ________. "Les enseignements de la guerre d'Espagne." L.A. 939 (June 15, 1939), 11.

502. ________. "Rechercher la supériorité aérienne, telle doit être la prochaine étape." L.A. 945 (July 27, 1939), 11.

503. ________. "Quelques aspects de la supériorité aérienne." L.A. 946 (Aug. 3, 1939), 11.

504. ________. "Quel avion majeur devons nous construire?"

L.A. 949 (Aug. 24, 1939), 11.

505. ________. "Pages d'histoire aéronautique: préliminaires de l'air intégral 1918-1926." L.A. 1696 (Sept. 27, 1958), 4, 5.

506. ________. "Pages d'histoire aéronautique: naissance et gloire de l'air intégral (1926-1930)." L.A. 1697 (Oct. 4, 1958), 4, 5.

507. ________. "Les incertitudes de l'air intégral (1926-1930)." L.A. 1698 (Oct. 11, 1958), 6, 7.

508. ________. "Mort de l'intégral et montée des périls." L.A. 1699 (Oct. 18, 1958), 4, 5.

509. Langevin, Commandant. "Action aérienne dans une bataille défensive." R.F.A. 6 (Jan., 1930), 21-64.

510. ________. "Les prémices de la 'chasse de nuit' en France." R.F.A. 9 (Apr., 1930), 400-416.

511. ________. "Combinaisons de monoplaces et de multiplaces." R.F.A. 27 (Oct., 1931), 1224-1230.

512. Laure, Général. LA DEFENSE AERIENNE DU TERRITOIRE. Paris: 1934.

513. La Vaulx, Comte Henry de. LES VAINQUEURS DE L'AIR. HISTOIRE DE L'AERONAUTIQUE: SES DEBUTS SPORTIFS, SON APPLICATION MILITAIRE. SA REALISATION COMMERCIALE. Paris: 1921.

514. Lebon, Emile. DE LA GUERRE AERIENNE DANS SES RAPPORTS AVEC LE DROIT INTERNATIONAL. Nancy: 1923.

515. Leclerc, Lieutenant-Colonel Rémy. "Ailes françaises en Tunisie." R.H.A. (Feb., 1955), 97-110.

516. Lecuir, J. and P. Fridenson. "L'organisation de la coopération aérienne franco-britannique (1935-mai 1940)." REVUE D'HISTOIRE DE LA DEUXIEME GUERRE MONDIALE 73 (Jan., 1969), 43-74.

517. Lefranc, J.A. "La tactique aérienne et la défense

contre avions." L'AEROPHILE, 1-2 (Jan. 1-15, 1919), 6-9; 7-8 (Apr. 1-15, 1919), 102-107; 11-12 (June 1-15, 1919), 161-168.

518. Le Goyet, Colonel. "Evolution de la doctrine d'emploi de l'aviation française entre 1919 et 1939." REVUE D'HISTOIRE DE LA DEUXIEME GUERRE MONDIALE 73 (Jan., 1969), 3-42.

519. Lehmann, O. "Die französische Luftfahrt in Spiegel der Zeit." DEUTSCHE WEHR I (1928), 408-409.

520. Lenoir, Maurice Lieutenant de Vaisseau. "Problèmes d'aviation d'escadre." R.A.A. 66 (Jan., 1935), 3-20.

521. _________. "Le Potez 63 peut être un bon avion d'arrêt." L.A. 831 (May 20, 1937), 13.

522. Le Wita, Henri. "Autour de la guerre chimique." L.A. 395 (Jan. 10, 1929), 5; 396 (Jan. 17, 1929), 5.

523. Lissarague, Général P. "Un constructeur, Henry Potez." R.H.A. (No. 2, 1977), 87-102.

524. Loges, Jean des. "L'aviation dans la bataille. La construction des forces aériennes." L.A. 325 (Sept. 8, 1927), 3.

525. _________. "L'aviation dans la bataille. De Napoléon a l'âge de l'Air." L.A. 336 (Nov. 24, 1927), 5.

526. "Limites d'utilisation des avions de chasse modernes." R.A.A. 75 (Oct., 1935), 1105-1117.

527. Logistica aerea nella campagna del Morocco (1925)." RIVISTA AERONAUTICA 3, No. 1 (1927), 69-73.

528. Luce-Catinot, Capitaine. "Essai sur l'utilisation de la radiotélégraphie dans l'armée de l'Air." R.A.A. 94 (May, 1937), 509-520.

529. Mackiewicz, Dymitr. "La tactique d'un avion biplace dans le combat avec les avions de chasse." R.A.A. 72 (July, 1935), 723-736.

530. Maincent, Paul. "Il ya 30 ans...la croisière Vuillemin

en Afrique noire (8 novembre-24 décembre 1933)." F.A.F. 198 (Dec., 1963), 787-804.

531. Mali, Jean. "Le personnel technique dans le ministère de l'Air." L.A. 384 (Oct. 25, 1928), 7.

532. ________. "Il materiale d'aviazione militare francese." RIVISTA AERONAUTICA 5 (Jan., 1929), 106-108.

533. ________. "L'unité technique dans le ministère de l'Air." L.A. 422 (July 18, 1929), 5.

534. ________. "La concentration industrielle." L.A. 448 (Jan. 16, 1930), 5, 6.

535. "Les maneuvres aériennes de 1924." R.A.M. 25 (Jan.-Feb., 1925), 1-4.

536A. Manuel, Jean. "L'emploi de la radio." L.A. 350 (Mar. 1, 1928), 5.

Marie, Colonel Félix. "Du tir à la bombe par avion." R.A.M. 18 (Nov.-Dec., 1923), 131-135.

537. Marie, Félix. "Importance du bombardement aérien dans la bataille." R.A.M. 32 (Mar.-Apr., 1926), 25-34.

538. ________. "L'aéronautique, arme savante." L.A. 529 (Aug. 6, 1931), 5.

539. "Le Maroc, école d'application." L.A. 214 (July 23, 1925), 2.

540. "Le Maroc et l'aviation. Le rôle actuel des avions; ce qu'il doit être demain." L.A. 220 (Sept. 3, 1925), 1.

541. "Le Maroc et l'aviation. Pour en finir vite et bien!" L.A. 214 (July 23, 1925), 1.

542. Martel, René. "L'histoire de l'aviation de bombardement." L.A. 848 (Sept. 16, 1937), 11.

543. Massenet de Marancour. LA CHASSE EN AVION. Paris: 1921.

544. Matignon, Capitaine. "La machine à finir la guerre." R.A.A. 78 (Jan., 1936), 3-29.

545. ________. "Tir aérien et méthodes de tir." R.A.A. 82 (May, 1936), 335.

546. Mattei, Jacques. "La recherche d'une doctrine. Comparaisons entre l'Air et la mer." L.A. 455 (Mar. 6, 1930), 5.

547. ________. "La 'récapitulation' du général Douhet." L.A. 458 (Mar. 27, 1930) 5, 6; 459 (Apr. 3, 1930), 5, 6.

548. Mangeri, Franco. "La dottrina Douhet in Francia." RIVISTA MARITTIMA (Sept., 1933), 97-102.

549. Mayer, Lieutenant-Colonel Emile. LA GUERRE D'HIER ET L'ARMEE DE DEMAIN. Paris: 1921.

550. ________. "La politique aérienne de la France. Guerre terrestre ou aéro-chimique?" L.A. 323 (Aug. 25, 1927), 3.

551. ________. "Les grandes controverses. L'aviation dans la guerre moderne." L.A. 351 (Mar. 8, 1928), 6.

552. Mecozzi, Amédeo. "L'opinion du Géneral Niessel devant l'aviation italienne." L.A. 395 (Jan. 10, 1929), 6, 7.

553. Melon, André. "L'entrainement des pilotes de réserves." L.A. 21 (Nov. 10, 1921), 4.

554. "La menace aérochimique. Comment Paris ou...Londres peuvent être détruits." I. L.A. 220 (Sept. 3, 1925), 1; II, 225 (Oct. 8, 1925), 1; III, 229 (Nov. 12, 1925), 1; IV, 234 (Dec. 10, 1925), 1; V, 239 (Jan. 14, 1926), 1; VI, 243 (Feb. 11, 1926), 1.

555. Meyer, Commandant André. AVOIR DES AILES. Préface du général Stehlin. Nancy: 1967.

556. Michelin, André. LA FRANCE A CREE L'AVIATION MAIS SI NOUS N'Y PRENONS GARDE, ELLE SERA DEMAIN UNE PUISSANCE AERIENNE DE TROISIEME ORDRE ET POURTANT

LA GUERRE AERIENNE RESTE LE GRAND DANGER DE DEMAIN. Paris: 1920.

557. ________. "Le facteur essentiel de la défense nationale." L.A. 340 (Dec. 22, 1927), 1.

558. ________. LE DANGER ALLEMAND. AVIATION ET GUERRE CHIMIQUE. Paris: 1922.

559. Michelin, André and Edouard. NOTRE SECURITE EST DANS L'AIR. Paris: 1919.

560. Milhaud, Edgard. "The Nationalization of the Aeronautical Industry in France and its Immediate Consequences." ANNALS OF COLLECTIVE ECONOMY 15 (1939), 223-251.

561. "Le Ministère de l'Air. Quelles sont les thèses en présence?" L.A. 375 (Aug. 24, 1928), 5, 6.

562. "Le Ministère de la Défense Nationale." R.A.A. 85 (Aug., 1936), 843-848.

563. Moineville, Commandant. "Les problèmes de la sureté dans la guerre aérienne autonome." R.M.A. 9 (Sept. 15, 1935), 1083-1109.

564. Molyneux, Paul. LA SANG DES AIGLES. Paris: 1930.

565. Monlau, Jacques. "L'Air et sa défense." L.A. 511 (Apr. 2, 1931), 5.

566. ________. "L'organisation de la défense aérienne." L.A. 576 (May 7, 1931), 5, 6.

567. ________. "Pour une école de l'air." R.F.A. 20 (Mar., 1931), 257-273

568. Montana, Pierre de. "Pour une école d'officier." L.A. 442 (Dec. 5, 1929), 5.

569. Montarby, Commandant de. "L'avion monomoteur est-il un avion de bombardement?" R.A.M. 35 (Mar.-Apr., 1927), 25-29.

570. Montaux, Pierre de. "Les difficultés de travail avec

les armées terrestres." L.A. 483 (Sept. 18, 1930),6.

571. ________. "L'armée ignore toujours l'aviation." L.A. 494 (Dec. 4, 1930), 14-15.

572. ________. "Une doctrine et un matériel." L.A. 495 (Dec. 14, 1930), 12, 13.

573. Morvan, André. "Sous les cocardes. Comme préface à une rubrique nouvelle." L.A. 285 (Dec. 3, 1926), 3.

574. ________. "Sous les cocardes. A propos d'une nomination." L.A. 286 (Dec. 9, 1926), 3.

575. ________. "Maroc et Syrie." L.A. 287 (Dec. 16, 1926), 3; 288 (Dec. 23, 1926), 3.

576. ________. "Un voyage rue Saint-Dominique." L.A. 291 (Jan. 13, 1927), 3.

577. ________. "Sous le signe du lecteur." L.A. 293 (Jan. 27, 1927), 3.

578. ________. "Le sous-officier dans l'aviation." L.A. 297 (Feb. 24, 1927), 3.

579. ________. "Une étude nécessaire: l'avion colonial." L.A. 298 (Mar. 3, 1927), 3.

580. ________. "Le statut de l'aéronautique militaire." L.A. 300 (Mar. 17, 1927), 3.

581. ________. "La maîtrise de l'air." L.A. 301 (Mar. 24, 1927), 3.

582. ________. "Le statut de l'aéronautique militaire." L.A. 305 (Apr. 21, 1927), 3; 306 (Apr. 28, 1927), 3; 307 (May 5, 1927), 3, 5; 308 (May 12, 1927), 3; 310 (May 26, 1927), 3; 312 (June 9, 1927), 3; 314 (June 23, 1927), 3.

583. ________. "L'aviation de nuit. Où en sommes nous?" L.A. 316 (July 7, 1927), 3.

584. ________. "L'avancement dans l'aviation." L.A. 318 (July 21, 1927), 3.

585. ________. "L'aviation dans la bataille. Qui a raison?..." L.A. 321 (Aug. 11, 1927), 3.

586. ________. "La chasse." L.A. 322 (Aug. 19, 1927), 3.

587. ________. "La chasse. Ses Moyens." L.A. 324 (Sept. 1, 1927), 3.

588. ________. "La sécurité dans l'aviation militaire. Trop d'accidents." L.A. 327 (Sept. 22, 1927), 3.

589. ________. "L'aviation légère. L'avion lèger est apte à la guerre." L.A. 337 (Dec. 1, 1927), 3.

590. ________. "Le statut de l'Aéronautique. Deux projets de loi." L.A. 339 (Dec. 15, 1927), 3.

591. ________. "La politique aérienne de la France. Les cadres et effectifs de l'Aéronautique." L.A. 344 (Jan. 19, 1928), 3.

592. ________. "La tenue de l'aviation. La conclusion d'une enquête." L.A. 346 (Feb. 2, 1928), 3.

593. ________. "La politique aérienne de la France. Les cadres et l'effectif aéronautique." L.A. 347 (Feb. 9, 1928), 5.

594. ________. "Les servitudes." L.A. 349 (Feb. 23, 1928), 5.

595. ________. "Le statut du personnel navigant. A propos de l'article 80." L.A. 351 (Mar. 8, 1928), 5.

596. ________. "La bataille continue..." L.A. 352 (Mar. 15, 1928), 5.

597. ________. "Une réforme nécessaire. Le statut du personnel." L.A. 360 (May 15, 1928), 5.

598. ________. "L'organe d'expériences tactiques." L.A. 361 (May 17, 1928), 5.

599. ________. "Le centre d'expériences tactiques." L.A. 365 (June 14, 1928), 5.

600. ________. "L'opinion du général Girod sur le

Ministère de l'Air." L.A. 368 (July 5, 1928), 16-17.

601. ________. "Le ministère de l'Air 'intégral' est une nécessité." L.A. 377 (Sept. 6, 1928), 10.

602. ________. "La rationalisation de l'aviation militaire." L.A. 378 (Sept. 13, 1928), 5; 379 (Sept. 20, 1928), 5.

603. ________. "En suivant les grandes manoeuvres aériennes." L.A. 380 (Sept. 27, 1928), 5.

604. ________. "Les aviateurs devant 'l'intégral'." L.A. 381 (Oct. 4, 1928), 5.

605. ________. "Le Plan du ministère de l'Air intégral." L.A. 380 (Sept. 27, 1928), 9, 10; 381 (Oct. 4, 1928), 9, 10; 382 (Oct. 22, 1928), 9, 10.

606. ________. "D.C.A. - Bombardement-Conseiller national." L.A. 383 (Oct. 18, 1928), 5.

607. ________. "Les marins devant 'l'intégral'." L.A. 384 (Oct. 25, 1928), 5.

608. ________. "Aviateur ou marin: il faut choisir." L.A. 386 (Nov. 8, 1928), 5.

609. M.R. "L'aviation en Indo-chine." L.A. 59 (Aug. 3, 1922), 4.

610. M.V. "Le grave danger. Peut-on parer à l'effroyable menace d'une guerre aéro-chimique?" L.A. 286 (Dec. 9, 1926), 4.

611. M.V. "La guerre aéro-chimique. La question des représailles et la société des nations." L.A. 290 (Jan. 6, 1927), 4.

612. Mysyrowicz, Ladislas. AUTOPSIE D'UNE DEFAITE. ORIGINE DE L'EFFONDREMENT MILITAIRE FRANCAIS DE 1940. Lausanne: 1973.

613. Narbonne, René de. LES AILES QUI POUSSENT. Paris: 1935.

614. Nebout, G.A. LE MINISTERE DE L'AIR. Preface by Colonel Brocard. Paris: 1929.

615. Néon. UNE ILLUSION. LA CONQUETE DE L'AIR. Paris: 1927.

616. "Das neugeschaffenen französischen Luftministerium." MILITAER-WOCHENBLATT Nov. 11, 1928, 694.

617. Niessel, Général A. LA MAITRISE DE L'AIR. Paris: 1928.

618. ________. PREPARONS LA DEFENSE ANTIAERIENNE. Paris: 1929.

619. ________. D.A.T. DEFENSE AERIENNE DU TERRITOIRE. Préface de Laurent-Eynac. Paris: 1934.

620. NNN. "Une stratégie nouvelle devant le danger aérien." L.A. 595 (Nov. 10, 1932), 5.

621. ________. "La guerre aérienne peut-elle être supprimée?" L.A. 597 (Nov. 24, 1932), 13.

622. ________. "L'équivoque de la guerre de l'air. De quoi s'agit-il?" L.A. 600 (Dec. 15, 1932), 6.

623. ________. "La mission des forces aériennes." L.A. 611 (Mar. 2, 1933), 13.

624. ________. "Les deux modes d'intervention du bombardement aérien." L.A. 622 (May 18, 1933), 7.

625. ________. "La dangereuse illusion du monoplace-canon. Ne renonçons pas, pour cette illusion, à la formule du multiplace qui, elle, est une réalité." L.A. 633 (Aug. 3, 1933), 13.

626. ________. "L'équivoque de l'armament aérien: de la controverse à l'expérience." L.A. 670 (Apr. 19, 1934), 6.

627. ________. "La critique du monoplace-canon." L.A. 673 (May 10, 1934), 12.

628. ________. "Pour un multiplace de combat a canon multitube." L.A. 674 (June 21, 1934), 11.

629. ________. "L'armée de l'Air agira-t-elle de jour?"

L.A. 691 (Sept. 13, 1934), 11.

630. "Note annexe sur l'effort de l'Etat de 1928 à 1933 pour la création d'avions et d'hydravions prototypes." R.A.A. 77 (Dec., 1935), 1370-1384.

631. "Note sur la participation de l'Etat à la création des prototypes d'avions, dans l'hypothèse d'une industrie à statut libre." R.A.A. 81 (Apr., 1936), 363-370.

632. Observer, P.C.C. "Les problèmes de la 'chasse'. Vitesse et armement." L.A. 437 (Oct. 31, 1929), 5; 439 (Nov. 19, 1929), 5, 6.

633. Odier, A. "La fermeture de la chasse." R.A.A. 86 (Sept., 1936), 969-970.

634. "L'offensive de M. Bouilloux Lafont. Une interview du rapporteur de la Commission des Finances." L.A. 71 (Oct. 26, 1922), 1.

635. "On France and World Politics." THE AEROPLANE, 29 (Sept. 2, 1925), 281-286.

636. "Les opérations aériennes au Maroc." R.A.M. 9 (May-June, 1922), 53-58; 10 (July-Aug., 1922), 83-86; 11 (Sept.-Oct., 1922), 111-114; 13 (Jan.-Feb., 1923), 10-12.

637. "Organisation de l'aéronautique militaire." R.A.M. 31 (Jan.-Feb., 1926), 1-4.

638. "L'organisation de la 5e arme." L.A. 47 (May 11, 1922), 1.

639. "L'organisation de notre industrie aéronautique sous le régime de la nationalisation." L.A. 814 (Jan. 21, 1937), 11.

640. "L'organisation des terrains d'aviation." R.A.M. 24 (Nov.-Dec., 1924), 131-138.

641. "Organización de la aeronautica maritima francesa." ICARO 50 (Feb.-Mar., 1932), 4-6, 9-13.

642. "L'organizzazione generale dell' aeronautica francese."

RIVISTA AERONAUTICA 3 (Mar., 1927), 75-77.

643. Orselli, capitaine. "Les nouveaux postes de TSF de l'aviation militaire." R.F.A. 7 (Feb., 1930), 200-209.

644. Orthlieb, Commandant. L'AERONAUTIQUE. HIER-DEMAIN. Paris: 1920.

645. "Où va l'aéronautique de coopération?" L.A. 647 (Nov. 9, 1933), 11.

646. Painlevé, Paul. DE LA SCIENCE A LA DEFENSE NATIONALE. DISCOURS ET FRAGMENTS. Paris: 1931.

647. Painlevé, Paul, Emile Borel and Charles Maurain. L'AVIATION. Paris: 1923.

Pacquier, Pierre. "Doctrine ou dogmatisme?" R.M.A. 5 (May, 1939), 586-594.

648. ________. GUERRE ET SURPRISE. Paris. n.d.

649. ________. "Guerre et surprise." R.M.A. 11 (Nov., 1935), 1423-1450; 12 (Dec. 15, 1935), 1490.

650. ________. "Guerre et vitesse." R.M.A. 23 (Nov. 15, 1936), 1445-1479.

651. ________. "Pensées du général Denain." F.A.F. 208 (Nov., 1964), 595-607.

652. ________. "La croisière noire." F.A.F. 242 (Dec., 1967), 877-910.

653. "Paris-Teheran-Paris." R.A.M. 33 (May-June, 1926), 62-65.

654. "La participation de l'aviation aux grandes manoeuvres anglaises de septembre, 1925." R.A.M. 30 (Nov.-Dec., 1925), 126-129.

655. "La participation de l'Etat à la création des prototypes d'avions." R.A.A. 77 (Dec., 1935), 1365-1369.

656. Pastier, Commandant. "L'expérience actuelle. Emploi de

l'aviation en liaison avec la cavalerie." R.A.M. 43 (Jan.-Feb., 1928), 14-16.

657. Pastier, Lieutenant-Colonel. "La question des réserves générales d'aviation." R.F.A. 33 (Apr., 1932), 363-374.

658. Pastier, Colonel. "Aviation lourde." R.M.A. 13 (Jan. 15, 1936), 5-13.

659. ________. "Aviation d'observation." R.M.A. 18 (June 15, 1936), 699-711.

660. Paulhan, Jeanne. LES AILES VICTORIEUSES. Paris: 1937.

661. Pennès Capitaine. "Réflexions sur l'emploi de l'aviation dans la réduction de la dissidence au sud Maroc." R.F.A. 7 (Feb., 1930), 186-199.

662. ________. "La première croisière aérienne d'escadre en Afrique française. Relation de voyage par le capitaine Pennès." R.A.A. 57 (Apr., 1934), 361-499.

663. Percheron, M. L'AVIATION FRANCAISE. Paris: 1929.

664. Perchon, Ernest. A L'OMBRE DES AILES. Paris: 1939.

665. Petit, Edmond. "Histoire des forces aériennes françaises, 1919-1939." ICARE 9/ (1981), 57-107.

666. ________. LA VIE QUOTIDIENNE DANS L'AVIATION EN FRANCE AU DEBUT DU XX SIECLE. 1900-1935. Paris: 1977.

667. Petitot, Capitaine. "Exploitation rapide du renseignement photographique aérien." R.F.A. 33 (Apr., 1933), 409-430.

668. Pfeiffer, Waldemar. "Formation de marche et action de l'aviation d'après les expériences de la guerre." R.A.M. 23 (Sept.-Oct., 1924), 107-109.

669. Picqué, R. Médecin. "L'Etat actuel de l'aviation sanitaire." R.A.M. 19 (Jan.-Feb., 1924), 15-20.

670. Pigeon, Capitaine. "Chasseurs contre bombardiers

modernes." R.A.A. 103 (Feb., 1938), 125-142.

671. ________. "Vol de groupe de bombardiers." R.A.A. 105 (Apr., 1938), 366-378.

672. ________. "Le bilan du chasseur." R.A.A. 117 (July-Aug., 1939) 375-409.

673. Un pilote. "L'aviation au Maroc: le bombardement? Oui! Le ravitaillement? Non!" L.A. 24 (Jan. 28, 1926), 3.

674. Piollet, Capitaine. "L'équipement terrestre de la manoeuvre aérienne." R.M.A. 9 (Sept. 15, 1935), 1126-1138.

675. "Politiques aériennes." L.A. 4 (July 14, 1921), 1.

676. Poncelet, L. "L'exploitation de la photographie aérienne." CONQUETE DE L'AIR 26 (May, 1930), 387-394.

677. Pouderoux, Général. "La chasse aérienne nocturne." R.A.A. 117 (July-Aug., 1939), 410-416.

678. Poulaine, Madeleine. "Les grandes figures de l'aviation: M. Etienne Riché." L.A. 578 (July 14, 1932), 8.

679. ________. "Les grandes figures de l'aviation: Laurent-Eynac." L.A. 599 (Dec. 8, 1932), 8.

680. Poulalion, Lieutenant de Vaisseau. "L'aviation coloniale." L.A. 40 (Mar. 23, 1922), 4; 41 (Mar. 30, 1922), 3; 42 (Apr. 6, 1922), 3.

681. "Project de loi fixant le statut organique de l'armée de l'Air. Rapport fait par M. Jacquinot, député, an nom de la Commission de l'Aéronautique." R.M.A. 2 (Feb., 1935), 213-221.

682. P., S. "La recherche d'une doctrine. Anticipation sur la bataille aérienne." L.A. 450 (Jan. 30, 1930), 5.

683. "Quel est l'observatoire volant idéal?" R.A.A. 113 (Dec., 1938) 1343-1345.

684. "Quelques opinions sur l'arme aérienne. Son emploi, ses possibilités." L.A. 214 (July 24, 1925), 1, 2.

685. Rabatel, Henri. "La technique et l'emploi du moteur-canon." L.A. 671 (Apr. 26, 1934), 7, 8.

686. ________. "La défense du monoplace-canon." L.A. 673 (May 10, 1934), 12.

687. ________. "Quelle est la meilleure solution? Le canon placé dans l'aile ou le moteur-canon?" L.A. /00bis (Nov. 18, 1934), 15.

688. Rabion, René. "L'Etat actuel de l'aviation russe." L.A. 229 (Nov. 5, 1925), 4.

689. "Le raid de Pelletier d'Oisy." L.A. 151 (May 8, 1924), 1; 152 (May 15, 1924), 1.

690. "Le rapport de M. Paul Rivers, député, sur le budget de l'air pour 1938." R.A.A. 103 (Feb., 1938), 181-198.

691. Renaitour, Jean Michel. LES FILS D'ICARE. HISTOIRES D'AVIATION. Paris: 1933.

692. "La réorganisation de l'Armée de l'Air." R.M.A. 22 (Oct. 15, 1936), 1231-1259.

693. Reveilhac, Jean. DIEUDONNE COSTES. Paris: 1983.

694. Riché, Etienne. "Une commission de l'aviation à la Chambre." L.A. 369 (July 13, 1928), 8.

695. ________. "Lettre ouverte à un collaborateur de la 'Revue des Deux mondes'." L.A. 386 (Nov. 8, 1928), 12.

696. ________. "Le ministère et l'armée de l'Air, organes nécessaires de la défense nationale." L.A. 403 (Mar. 7, 1929), 3, 4.

697. ________. "Le statut du personnel de l'aéronautique nationale." L.A. 469 (June 12, 1930), 5.

698. ________. "Les forces aériennes devant le Parlement.

I - Les forces aeriennes et la politique de prix." L.A. 470 (June 19, 1930), 7, 8; "II - L'organisation générale aérienne" 471 (June 26, 1930), 6; "III - Le commandement et l'organisation technique" 472 (July 3, 1930), 6, 7; "IV - Le progrès tactique et la mobilisation" 473 (July 10, 1930), 6, 7; "V - La formation des cadres" 474 (July 17, 1930), 6, 7.

699. ________. "L'organisation de l'Etat et le danger aérien." L.A. 484 (Sept. 25, 1930), 5, 6.

700. ________. "A la recherche d'une loi." L.A. 578 (July 14, 1932), 5.

701. ________. "La sécurité française devant le désarmement aérien." L.A. 599 (Dec. 8, 1932), 5, 6.

702. Rignot, Capitaine. "Les grands voyages aériens. De Paris en Sibérie." R.A.M. 40 (July-Aug., 1927), 82-85.

703. Ritter, Hans. "Post-War Fighter Progress." AVIATION 30 (Dec., 1931), 692-694.

704. Ritter, Capitaine D. "La guerre aérienne doit-elle être autonome ou non?" R.F.A. 49 (Aug., 1933), 861-880.

705. Rives, Paul. "Le personnel de l'armée de l'Air." R.A.A. 116 (May-June, 1939), 269-290.

706. Rodet, Capitaine Paul. "L'aviation d'artillerie." R.A.A. 119 (Nov.-Dec., 1939), 627-632.

707. Romatet, Lieutenant J. "Organisation de l'aéronautique militaire." R.A.M. 1 (Jan.-Feb., 1921), 3-6.

708. ________. "L'avion de chasse. Son passé, son avenir. L'expérience de la guerre." R.A.M. 2 (Mar.-Apr., 1921), 32-37.

709. Romatet, Commandant. "Monoplaces et multiplaces de chasse." R.F.A. 29 (Dec., 1931), 1412-1419.

710. Romeyer, Jean. "Un général italien affirme sa foi dans

l'armée de l'Air." L.A. 352 (Mar. 15, 1928), 4.

711. ________. "La menace aéro-chimique. Considérations objectives sur la réalité de cette menace." L.A. 399 (Feb. 7, 1929), 8-10.

712. Roques. "L'aéronautique coloniale." R.F.A. 1 (Aug., 1929), 65-83.

713. Rossi. AU SERVICE DE L'AVIATION FRANCAISE (1919-1939). Paris: 1941.

714. Roubleau, H. "L'aviation aux aviateurs." L.A. 45 (Apr. 27, 1922), 1.

715. Rougeron, C. "L'aviation de bataille." R.F.A. 19 (Feb., 1931), 155-168.

716. ________. "L'attaque des flottes." L.A. 507 (Mar. 5, 1931), 5.

717. ________. "Le cuirasse et l'avion." L.A. 509 (Mar. 19, 1931), 5.

718. ________. "Controverse aéro-maritime." L.A. 512 (Apr. 9, 1931), 5.

719. ________. "La guerre de l'Air. Examen des thèses classiques." L.A. 518 (May 21, 1931), 5.

720. ________. "Le rendement du bombardement aérien." L.A. 519 (May 28, 1931), 5, 6.

721. ________. "L'efficacité des méthodes actuelles." L.A. 520 (June 4, 1931), 5.

722. ________. "La destruction des villes." L.A. 537 (Oct. 1, 1931), 5.

723. ________. "La guerre de l'Air. L'action contre les forces organisées." L.A. 545 (Nov. 26, 1931), 5, 6.

724. ________. "Les réserves générales d'aviation." R.F.A. 45 (Apr., 1933), 363-381.

725. ________. "Douhet et la doctrine de l'unité

matériel." R.A.A. 58 (May, 1934), 503-536.

726. ________. "L'efficacité du bombardement aérien." R.A.A. 64 (Nov., 1934), 1227-1246.

727. ________. "Vers l'avion de chasse, avion unique." L.A. 703 (Dec. 6, 1934), 10-11.

728. ________. "Le bombardement en piqué." R.A.A. 65 (Dec., 1934), 1327-1416.

729. ________. "Le bombardement des gares et des voies ferrées." R.A.A. 72 (July, 1935), 723.

730. ________. L'AVIATION DE BOMBARDEMENT. 2 vols. Paris: 1936.

731. ________. "Le bombardement des immeubles d'habitations." R.A.A. 83 (June, 1936), 661-674.

732. ________. "Le bombardement en vol rasant." R.A.A. 86 (Sept., 1936), 1007-1024.

733. ________. "L'aviation et les chars de combat." R.A.A. 92 (Mar., 1937), 249-285.

734. ________. "L'aviation sur l'échiquier européen." R.A.A. 96 (July, 1937), 731-745.

735. ________. "La défense de l'Afrique du nord ou la defense par l'Afrique du nord?" R.A.A. 101 (Dec., 1937), 1328-1336.

736. ________. "Les aspects militaires du trafic aérien commercial. Les transports aériens en temps du guerre." R.A.A. 105 (Apr., 1938), 379-397.

737. ________. LES ENSEIGNEMENTS AERIENS DE LA GUERRE D'ESPAGNE. Paris: 1939.

738. ________. "L'attaque en vol rasant entre les navires marchands." L.A. 928 (Mar. 30, 1939), 11.

739. ________. "La défense des navires marchands contre les attaques aériennes." L.A. 930 (Apr. 13, 1939), 11.

740. ________. "La destruction de l'aviation à terre." R.A.A. 115 (Mar.-Apr., 1939), 131-148.

741. Saint-Quentin, G. de. LE GUERRE DE 1924. Saint Cloud les Coteaux: 1923.

742. Sclafer, Commandant. "L'aviation militaire en Mauritanie." R.F.A. 13 (Aug., 1939), 927-946.

743. Schlumberger, Chef de bataillon. "Evolution des avions de coopération." R.A.A. 94 (May, 1937), 521-529.

744. Seive, Général. "Lyon, centre d'aviation militaire." R.H.A. 2 (May, 1958), 143-158.

745. Senille, Capitaine. "Rénovation du matériel d'aérostation." R.A.A. 62 (Sept., 1934), 1020-1030.

746. Serre, Captaine de frégate H. "Emploi de l'aviation dans la guerre navale." R.F.A. 24 (July, 1931), 839-868.

747. ________. "L'offensive aérienne dans la guerre sur mer." R.F.A. 35 (June, 1932), 603-632.

748. Seyffardt, H.A. "Die Lage der französischen Luftfahrt und das neue Luftministerium." DIE LUFTWACHT 2 (Feb., 1929), 62-65; 5 (May, 1929), 200-204.

749. "Si la coopération n'avait pas existé, il aurait fallu l'inventer." R.A.A. 102 (Jan., 1938), 3-6.

750. Simon, Louis, and Maurice Arnoux. DEFENSE PASSIVE CONTRE LES ATTAQUES AERIENNES. Preface by Général Duchène. Paris: 1936.

751. Simon-Fromant. "L'armée de l'Air indépendante. Le facteur essentiel de la défense nationale." L.A. 568 (May 5, 1932), 5.

752. Simoune, René. "Le danger aérien. L'inquiètude de l'Europe." L.A. 513 (Apr. 16, 1931), 5, 6.

753. Skidmore, Ellen T. "Pierre Cot: Apostle of Collective Security, 1919-1939." Doctoral Dissertation, University of Tennessee, 1980.

754. Sologne, René. "L'aviation de Douhet et la nôtre." L.A. 574 (June 16, 1932), 5.

755. Sorb, Commandant. "La marine et l'aviation." L.A. 558 (Feb. 25, 1932), 5.

756. Sorlin, Pierre. "Les perspectives aéronautiques dans la crise tchécoslovaque de 1938." F.A.F. 142 (Nov., 1958), 601-636.

757. Spivak, Marcel. "Les problèmes posés à l'armée de terre par la création du ministère de l'air et les perspectives d'indépendance de l'armée de l'Air (1928-1934)." COLLOQUE INTERNATIONALE. ADAPTATION DE L'ARMEE AERIENNE AUX CONFLITS CONTEMPORAINS ET PROCESSUS D'INDEPENDANCE DES ARMEES DE L'AIR DES ORIGINES A LA FIN DE LA SECONDE GUERRE MONDIALE. Paris: 1985, 177-189.

758. Stehlin, Lieutenant. "Le potentiel militaire d'une aviation commerciale: la 'Deutsche Lufthansa'." R.M.A. 3 (Mar., 1935), 268-284.

759. Stehlin, Paul. TEMOIGNAGE POUR L'HISTOIRE. Paris: 1964.

760. Stewart, Oliver. "The Air Forces of France." ROYAL UNITED SERVICE INSTITUTE JOURNAL 79 (1934), 769-779.

761. Stiot, Commissaire Général. "L'aviation militaire, ses premiers uniformes." UNIFORMES (No. 60, 1981), 35-41; (No. 61, 1981), 8-12.

762. Suarnet, André. LES RAMPANTS. SOUVENIRS D'ESCADRILLES. Paris: 1930.

763. "Sur la surprise aéronautique." R.A.A. 119 (Nov.-Dec., 1939), 603-606.

764. "Sur une aviation d'artillerie autonome." R.A.A. 114 (Jan.-Feb., 1939), 3, 4.

765. Thédenat, Lieutenant de Vaisseau. "Avions bombardiers et navires de guerre." R.M.A. 6 (June 15, 1935), 699-717.

766. Thomas, M.R. "Les premiers raids aériens en Afrique (1911-1933)." F.A.F. 165 (Dec., 1960), 739-768.

767. Thoumin, Capitaine R. "Recherche du champ de bataille." R.A.A. 94 (May, 1937), 483-499.

768. Thoumin, Captaine, R. "Formation de l'observateur." R.A.A. 102 (Jan., 1938), 21-40.

769. Toussaint, A. L'AVIATION ACTUELLE. Paris: 1928.

770. LA TRAHISON. Preface by J. Doriot. Paris: 1942.

771. "Travail de l'Aéronautique d'Indochine." R.A.M. 27 (May-June, 1925), 56-61.

772. Truelle, Ingénieur Général J. "La production aéronautique militaire française jusqu'en juin 1940." REVUE D'HISTOIRE DE LA DEUXIEME GUERRE MONDIALE 73 (Jan., 1969), 75-109.

773. "Utilisation des photographies aériennes pour les levers topographiques de Sedan et de Tarbes au 10.000°." R.A.M. 24 (Nov.-Dec., 1924), 139-142.

774. Vaïsse, Maurice. "Le procès de l'aviation de bombardement." R.H.A. (No. 2, 1977), 41-61.

775. Vauthier, Commandant P. QUESTIONS D'ARTILLERIE ANTIAERIENNE. Paris: 1928.

776. Vauthier, Lieutenant-Colonel. LA DEFENSE ANTI-AERIENNE DES GRANDES UNITES. Paris: 1929.

777. ________. LE DANGER AERIEN ET L'AVENIR DU PAYS. Preface by Maréchal Lyautey. Paris: 1930.

778. ________. "Les detachements armés, transportés par avions." R.F.A. 12 (July, 1930), 804-820.

779. ________. "Le danger aérien et l'avenir du pays." L.A. 487 (Oct. 16, 1930), 6, 7.

780. Vauthier, Colonel P. LA DOCTRINE DE GUERRE DU GENERAL DOUHET. Preface by Maréchal Pétain. Paris: 1935.

781. ________. "Principes de la doctrine de Douhet."

R.A.A. 69 (Apr., 1935), 363-382.

782. _________. "Vitesse, armament, autonomie." R.M.A. 8 (Aug. 15, 1935), 951-971.

783. _________. "A propos de Douhet." R.M.A. 15 (Mar. 15, 1936), 365-371.

784. Verax. "Un grand procès. 1 - L'opinion doit intervenir daus un début grave." L.A. 403 (Mar. 7, 1929), 5; "2 - Considerations sur l'envie et la jalousie." 404 (Mar. 14, 1929), 5; "3 - Lettre ouverte aux parlementaires." 405 (Mar. 21, 1929), 5; "4 - L'impossibilité du possible." 406 (Mar. 28, 1929), 5; "5 - considerations sur l'unité du commandement." 407 (Apr. 4, 1929), 5.

785. _________. "Premier regard sur un projet de loi." L.A. 417 (June 13, 1929), 5.

786. _________. "Le ministère de l'Air ne doit pas être qu'un magasin." L.A. 418 (June 20, 1929), 5.

787. _________. "1930. Explication du temps présent." L.A. 448 (Jan. 16, 1930), 5.

788. _________. "Une erreur, une hérésie." L.A. 454 (Feb. 27, 1930), 5.

789. _________. "Les aviateurs devant l'opinion." L.A. 458 (Mar. 27, 1930), 5.

790. _________. "Sur le conseil supérieur de l'Air." L.A. 534 (Sept. 10, 1931), 5.

791. Verdurand, Lieutenant-Colonel. "Une question simple: l'armée de l'Air doit-elle être la seule à possèder des avions?" R.A.A. 115 (Mar.-Apr., 1939), 195-197.

792. Vincens, Commandant. "L'expérience actuelle. L'attaque de nuit d'un objectif très défendu." R.A.M. 45 (May-June, 1928), 65-66.

793. Voisin, Général. "Monoplaces ou biplaces de chasse?" R.F.A. 30 (Jan., 1932), 42-67.

794. Vouillemin, Général. "La défense contre les avions. (D.C.A.)." R.A.M. 39 (May-June, 1927), 49-52

795. "Le voyage Paris-Addis Abeba d'un équipage militaire français." R.F.A. 21 (Apr., 1931), 454-467.

796. Vuillemin, Général. "Confiance et cohésion." L.A. 911 (Dec. 1, 1938), 21.

797. Weiss, Commandant Pierre. LES CHARMEURS DE NUAGE. Paris: 1928.

798. Weiss, Pierre. ESCALES ET PAYSAGES. Paris: 1939.

799. Wing. "Les marchés de l'état." L.A. 76 (Nov. 30, 1922), 4.

800. Woelflin, Capitaine. "L'emploi de l'aviation tel qu'il est conçu par les allemands." R.A.M. 20 (Mar.-Apr., 1924), 41-44; 21 (May-June, 1924), 60-62.

801. ________. "Utilisation de l'avion pour la création de masques de fumée et la projection de produits toxiques." R.A.M. 41 (Sept.-Oct., 1927), 113-114.

802. ________. "L'aviation et la guerre chimique." R.A.M. 46 (July-Aug., 1928), 95-101.

803. X... Commandant. "L'aviation en guerre de montagne." R.F.A. 2 (Sept., 1929), 173-193.

804. X. "Ait-Yacoub. Le rôle de l'aviation dans les affairs de Guefifat, Tarda et Ait-Yacoub au Maroc." R.F.A. 3 (Oct., 1929), 295-308.

805. ________. "La doctrine d'emploi de l'armée de l'Air." R.M.A. 8 (Aug. 15, 1935), 1006-1013; 9 (Sept. 15, 1935), 1051-1058.

806. XX. "La véritable naissance de l'école de l'Air." F.A.F. 154 (Dec., 1959), 1103-1110.

807. XXX. "La maitrise de l'air." L.A. 20 (Nov. 3, 1921), 3.

808. ________. "Le danger russe." L.A. 89 (Mar. 1, 1923),

1, 4.

809. ________. "Les Ailes et la guerre. Les sages paroles de M.P.E. Flandin." L.A. 95 (Mar. 12, 1923), 4.

810. ________. "La politique aérienne de la France. Réflexions sur le temps présent." L.A. 329 (Oct. 6, 1927), 3.

811. ________. "Réflexions sur le temps présent: l'aviation de renseignement." L.A. 332 (Oct. 27, 1927), 3.

812. ________. "Réflexions sur le temps présent. L'aviation de renseignement. La confusion des missions." L.A. 335 (Nov. 17, 1927), 3; 341 (Dec. 29, 1927), 3; 343 (Jan. 12, 1928), 3.

813. ________. "Les manoeuvres aériennes françaises." L.A. 379 (Sept. 20, 1928), 9, 10.

814. ________. "L'organisation tactique de l'armée aérienne." L.A. 387 (Nov. 15, 1928), 5.

815. ________. "Réflexions sur l'aviation de chasse monoplace." L.A. 393 (Dec. 27, 1928), 5.

816. ________. "La recherche d'une doctrine. L'aviation de renseignement dans l'armée de l'Air." L.A. 397 (Jan. 24, 1929), 5.

817. ________. "L'organisation des régiments d'aviation." L.A. 399 (Feb. 7, 1929), 6, 7.

818. ________. "Ce que doit être l'avion monoplace de chasse." L.A. 441 (Nov. 28, 1929), 5, 7.

819. ________. "Les sources de la victoire aérienne." L.A. 486 (Oct. 9, 1930), 6, 7.

820. ________. "Le problème des armements aériens." L.A. 615 (Mar. 30, 1933), 2.

821. Young, R.J. "The Strategic Dream: French air doctrine in the interwar period: 1919-1939." JOURNAL OF CONTEMPORARY HISTORY 9 (Oct., 1974), 57-76.

822. ________. "Le haut commandement français au moment de Munich." R.H.M.C. 24 (Jan.-Mar., 1977), 110-129.

823. ________. IN COMMAND OF FRANCE. FRENCH FOREIGN POLICY AND MILITARY PLANNING: 1933-1940. London: 1978.

824. Z... "Aéronautique militaire et aéronautique civile." R.A.M. 10 (July-Aug., 1922), 73-75.

825. ZZZ. "Des avions coloniaux sont à créer. Il faut y songer." L.A. 426 (Aug. 22, 1929), 5.

Here again, it is the broad, general studies which are still lacking. A good starting place might be the outline history of the Armée de l'Air from 1939 to 1945 by Patrice Buffotot and Patrick Facon (67); also useful are the books by Pierre Pacquier and his collaborators in the series LES FORCES AERIENNES FRANCAISES DE 1939 A 1945 (224, 225, 257) though these are now somewhat dated and are more narrative than explanatory in tone. The series of articles in several numbers of ICARE which are dedicated to this period are well worth exploiting (1-10), and also quite useful for the fighter arm is the study by Jean Gisclon (133).

The defeat of 1940 has been the subject of a number of works though the greater part of them are content simply to relate events. A good point of departure is the excellent statistical and analytical effort undertaken by Patrice Buffotot and P. Ogier (68). This can be supplemented with J. Accart's ON S'EST BATTU DANS LE CIEL (11) and General d'Astier de la Vigerie's LE CIEL N'ETAIT PAS VIDE (22), and by Jean Gisclon's ILS OUVRIRENT LE BAL (132). Eddy Bauer's essay on the air battles of May-June, 1940 (34) is excellent, particularly when coupled with Marc Bloch's magnificent analysis (42), which reveals much about the French military mentality in general during this period. The basic study of the aeronautical industry is that of General Christienne (75).

Work on the air forces under the armistice of 1940 was barely begun, so writings that can be cited are few. There is, however, an article by Patric Facon and Françoise de Ruffray on aeronautical collaboration with the Germans in 1940-1943 (121).

The Free French Air Forces still have not been extensively studied, at least not by serious scholars; there have not been any theses on the subject save one which remains incompleted. Most of the publications that have appeared have been memoirs or essentially narrative histories; among the better first-hand accounts one might cite those of General Andrieux (18), F. Broche (52), Pierre Clostermann (78, 79) and Bernard Duperier (100). Also worthwhile are the diaries of René Mouchotte (211, 212) and General Valin's LES SANS-CULOTTES DE L'AIR (274).

The essential works covering the rearming of the Armée de l'Air in North Africa and its participation in the war on the Allied side are those of General Bouscat (50) and Marcel Vigueras' REARMING THE FRENCH (277). Also useful is a series of articles by Patric Facon dealing with rearmament in 1942-43 (112), Plan VII (113), Franco-Allied cooperation in the Tunisian campaign (118), and the problems involved in returning the Armée de l'Air to France (120). Also useful though more specialized are A. Thibaudet's study of the rebirth of the French fighter arm (270) and Commissaire Mondins work on rearmament in North Africa in 1943 (210). Lastly the question of the Resistance within the Armée de l'Air is treated in J. Bezy's book LE S.R. AIR (40) and in A. Dainville's L'ORA (90).

BIBLIOGRAPHICAL LISTING

1. "1939-1940. La drôle de guerre." ICARE 53 (1970), 22-193.

2. "1939-1940. La bataille de France. Volume I: la chasse." ICARE 54 (1970), 34-139.

3. "1939-1940. La bataille de France. Volume II: la chasse." ICARE 55 (1970), 34-141.

4. "1939-1940. La bataille de France. Volume III: le bombardement, la reconnaissance." ICARE 57 (1971), 26-155.

5. "1939-1940. La bataille de France. Volume IV: la reconnaissance et les groupes aériens d'observation." ICARE 59 (1971), 43-179.

6. "1939-1940. La bataille de France. Volume V: L'aéronavale, première partie." ICARE 60 (1972), 34-137.

7. "1939-1940. La bataille de France. Volume VI: L'aéronavale, deuxième partie." ICARE 61 (1972), 33-145.

8. "1939-1940. La bataille de France. Volume X: l'aviation d'assaut, la 51e escadre." ICARE 80 (1977), 2-89.

9. "1939-1940. La bataille de France. Volume XI: l'aviation d'assaut, la 54e escadre." ICARE 87 (1978-1979), 2-83.

10. "1939-1940. La bataille de france. Volume XII: l'aviation d'assaut, la 54e escadre." ICARE 94 (1980), 2-93.

11. Accart, Capitaine Jean. ON S'EST BATTU DANS LE CIEL. Paris: 1942.

12. _________. LE PILOTE SOLITAIRE. Grenoble: 1944.

13. _________. EVADES DE FRANCE. Paris: 1945.

14. _________. CHASSEURS DU CIEL. Paris: 1946.

15. _________. SILENCE, ON VOLE. Paris: 1946.

16. Alerme, Colonel. LES CAUSES MILITAIRES DE NOTRE DEFAITE. Paris: 1940.

17. Alexander, Martin S. "Force de frappe ou feu de paille. Maurice Gamelin's Appraisal of Military Aviation before the Blitzkrieg of 1940." COLLOQUE INTERNATIONAL: L'ADAPTATION DE L'ARME AERIENNE AUX CONFLITS CONTEMPORAINS ET PROCESSUS D'INDEPENDANCE DES ARMEES DE L'AIR. Paris: 1985.

18. Andrieux, Colonel Jacques. LE CIEL ET L'ENFER, FRANCE LIBRE 1940-1945. Paris: 1965.

19. _________. UNE POIGNEE D'AS. Paris: 1976.

20. Anthérieu, Etienne. LE DRAME DE L'ARMEE D'ARMISTICE. Paris: 1946.

21. Aron, Robert. BOMBARDIERS DE NUIT, LES GROUPES LOURDS SUR L'ALLEMAGNE. Paris: 1946.

22. Astier de la Vigerie Général d'. LE CIEL N'ETAIT PAS VIDE. Paris: 1952.

23. Auber, A. JANICK, CAPITAINE AU GROUPE DE CHASSE NAVARRE, 1918-1944. Paris: 1946.

24. AVIATION DE FRANCE. Vichy: 1944.

25. L'AVIATION FRANCAISE. Lyon: 1942.

26. L'AVIATION FRANCAISE AU COMBAT SUR LE FRONT DE L'OUEST: CHASSEURS BOMBARDIERS. Paris: 1946.

27. Baitson, Robert. L'OMBRE SUR LES COCARDES. Lyon: 1945.

28. Ballande, Henri. DE L'AMIRAUTE A BIKINI. SOUVENIRS DE JOURS SANS GLOIRE. Paris: 1972.

29. Ballanger, Bernard. "Bombardement aérien de la ville de Royan (Charente-Maritime) le 5 janvier 1945." Unpublished thesis, doctorat d'université, Université de Bordeaux III, 1978.

30. Baquet, Michel. LES AIGLES DE CAMBRAI. Cambrai: 1979.

31. Baradez, Jean. AVIATEUR. Paris: 1940.

32. Barthélémy, Général. L'ESCADRILLE DU CALAO. Paris: 1976.

33. Bataille, Colonel Jean. UN ELEVE-PILOTE AUX ETATS-UNIS. Paris: 1946.

34. Bauer, Eddy. "Opinions sur la bataille aérienne de mai-juin 1940." F.A.F. 119 (Oct., 1956), 695-706.

35. Beaufre, Général. LE DRAME DE 1940. Paris: 1965.

36. Béguin, Capitaine. "Normandie en Russie." BULLETIN DES FORCES AERIENNES FRANCAISES EN GRANDE-BRETAGNE 6 (1944), 3-5.

37. Belugou, Sylvie. UN AIGLE DU I/2. n.p.: 1970.

38. Berge, Colonel. "Histoire des parachutistes de la France libre." F.A.F. 7 (Apr., 1947), 64-75.

39. Bernet, Robert. MON CARNET DE VOL 1940. Troyes: 1950.

40. Bezy, Jean. LE S.R. AIR. Paris: 1979.

41. Bisson, Colonel. LES OPERATIONS AERIENNES PENDANT LA CAMPAGNE DE 1939-1940. Paris: 1947.

42. Bloch, Marc. L'ETRANGE DEFAITE. Paris: 1946.

43. Blot, Camille. "Les opérations aériennes britanniques pendant la campagne de France." F.A.F. 99 (Dec., 1954), 953-978.

44. Bodet, Général P. "Le réarmement du bombardement français moyen sur Marauder, 1943-1945." F.A.F. 258 (May, 1969), 555-566.

45. Botquin, Gaston. "The Dewoitine D.520." A.H. 25 (Fall, 1978), 153-154.

46. Bougerol, Guy. CEUX QU'ON A JAMAIS VUS. Grenoble: 1943.

47. Bourgain, Louis. SARABANDE NOCTURNE. Nancy: 1951.

48. Bouscat, Général René. "L'armée de l'air française dans la campagne d'Italie." R.D.N. Feb., 1946, 233-237.

49. ________. "Marin La Meslée." F.A.F. 29 (Feb., 1949), 679-681.

50. ________. DE GAULLE-GIRAUD: DOSSIER D'UNE MISSION. Paris: 1967.

51. Briand, Guy. "L'autogire dans la guerre en 1940." L.A. 1151 (Feb. 21, 1948), 5-6.

52. Broche, François. LES BOMBARDIERS DE LA FRANCE LIBRE. Paris: 1979.

53. Buffotot, Patrice. "L'extraordinaire exploit d'un équipage du groupe Franche-Comté abattu le 19 août 1944 au-dessus de Toulon." R.H.A. (No. 4, 1974), 232-241.

54. ________. "Le moral dans l'armée de l'air française de septembre 1939 à juin 1940." FRANCAIS ET BRITANNIQUES DANS LA DROLE DE GUERRE. Paris: 1979, 173-196.

________. "L'armée de l'air française en Indochine (1940-1945)." M.A. 16 (Jan., 1974), 35-41.

________. "L'armée de l'air française pendant la drôle de guerre (3 septembre 1939-9 mai 1940)." M.A. 18 (Mar., 1979), 22-29.

55. ________. "L'armée de l'air pendant la bataille de 1940: la fin des combats et le bilan." M.A. 20 (May, 1979), 23-31.

56. ________. "Les unités polonaises de l'armée de l'air." M.A. 21 (June, 1979), 10-16.

57. ________. "Les missions de bombardement au-dessus de l'Allemagne à travers le journal de marche du groupe de bombardement I/22 Maroc, du ler janvier au 8 mai 1945." R.H.A. (No. 2, 1979), 154-168.

58. ________. "Le projet de bombardement des pétroles soviétiques du Caucase en 1940." R.H.A. (No. 4, 1979), 79-101.

59. ________. "Les forces aériennes françaises libres en Afrique: l'escadrille Topic du GRB.1 à Koufra en février 1941." R.H.A. (No. 2, 1980), 171-188.

60. ________. "L'aide aérienne à la Finlande." R.H.A. (No. 4, 1980), 175-200.

61. Buffotot, Patrice and Patrick Facon. "La naissance des forces aériennes françaises libres (juin 1940-juillet 1941)." M.A. 9 (June, 1978), 19-25.

62. ________. "L'affirmation des forces aériennes françaises libres (juillet 1941-août 1943)." M.A. 10 (July, 1978), 18-23.

63. ________. "La renaissance de l'armée de l'air française: novembre 1942-juin 1944." M.A. 11 (Aug., 1978), 27-32.

64. ________. "L'armée de l'air française dans la victoire." M.A. 12 (Sept., 1978), 22-31.

65. ________. "L'armée de l'air d'armistice: de l'affaire de Mers el-Kébir aux opérations du Levant (juillet 1940-juillet 1941)." M.A. 13 (Oct., 1978), 23-30.

66. ________. "L'armée de l'air de Vichy et l'affaire de Syrie (mai-juillet 1941)." M.A. 14 (Nov., 1978), 28-34.

67. ________. HISTOIRE DE L'ARMEE DE L'AIR FRANCAISE DE 1939 à 1945. Neuilly: 1980.

68. Buffotot, Patrice and Jacques Odier. "L'armée de l'air pendant la bataille de France (du 10 mai à l'armistice). Essai de bilan numérique d'une bataille aérienne." R.H.A. (No. 3, 1975), 88-117.

69. Calmel, Jean. PILOTES DU NUIT. Paris: 1952

70. ________. NIGHT PILOT. London: 1955.

71. C.G.D.A.: GROUPEMENT DE CHASSE ET DE DEFENSE AERIENNE. Paris: 1946.

72. Chambe, Général René. EQUIPAGES DANS LA FOURNAISE. Paris: 1945.

73. Charles, Lieutenant-Colonel R. "Les forces aériennes françaises libres en Afrique." R.H.A. (No. 4, 1964), 101-126.

74. Chassin, Général L.M. "Un plan grandiose: l'attaque des pétroles du Caucase en 1940." F.A.F. 176 (Dec., 1961), 821-850.

75. Christienne, Général Charles. "L'industrie aéronautique française de septembre 1939 à juin 1940." FRANCAIS ET BRITANNIQUES DANS LA DROLE DE GUERRE. Paris: 1979, 389-410

76. _________. "Les ailes françaises dans le ciel au jour J." RECUEIL D'ARTICLES ET ETUDES (1974-1975). Vincennes: 227-238.

77. DANS TOUS LES CIELS DE GUERRE: LA FRANCE. Paris: 1945.

78. Clostermann, Pierre. LE GRAND CIRQUE. Paris: 1948.

79. _________. THE BIG SHOW: SOME EXPERIENCES OF A FRENCH FIGHTER PILOT IN THE R.A.F. Translated by Oliver Berthoud. London: 1951.

80. _________. FEUX DU CIEL. Paris: 1951.

81. _________. FLAMES IN THE SKY. Translated by Oliver Berthoud. London: 1952.

82. Codos, Paul. ROUTES DU CIEL. Paris: 1955.

83. Constantini, Colonel A. "Le régiment Normandie-Niemen dans l'ultime phase de la campagne de Prusse Orientale (avril 1945)." R.H.A. (No. 4, 1977), 115-124.

84. _________. "Le groupe Normandie-Niemen dans la bataille de Koursk." R.H.A. (No. 3, 1973), 119-130.

85. Corniglion-Molinier, Colonel E. "Partant pour la Russie." BULLETIN DES FORCES AERIENNES FRANCAISES EN GRANDE-BRETAGNE 1 (1944), 3-6.

86. Cot, Pierre. LE PROCES DE LA REPUBLIQUE. 2 vols. New York: 1944.

87. ________. THE TRIUMPH OF TREASON. Chicago: 1944.

88. Courrière, Yves. NORMANDIE-NIEMEN. UN TEMPS POUR LA GUERRE. Paris: 1979.

89. Cuny, J. "Armée de l'Air in the European Theatre on May 10, 1940." R.A.F. FLYING REVIEW 10 (1958), 13-16.

90. Dainville, Colonel A. de. L'ORA, LA RESISTANCE DE L'ARMEE 1939-1945. Paris: 1974.

91. Delaunay, Henri. ARAIGNEE DU SOIR. Paris: 1968.

92. "Le débarquement. La chasse (Corse, Normandie, Provence)." ICARE 109 (1984), 7-153.

93. "Le débarquement. Le Groupe Lorraine et Alsace." ICARE 110 (1984), 7-153.

94. "Le débarquement. Groupes de bombardement et de reconnaissance." ICARE 111 (1984), 6-147.

95. Decot, Daniel. PILOTES FRANCAIS SUR LA VALLEE DU RHONE. Paris: 1982.

96. Destieu, Jean. CHASSEUR DE LA FRANCE LIBRE. Sermeac: 1978.

97. Dignac, Pierre. LES MALFAITEURS PUBLICS. Grenoble: 1941.

98. Dormann, Maurice. LA RADIEUSE EPOPEE DE MAURICE ARNOUX. Paris: 1946.

99. Doucet, Daniel. "Le Général Valin et les FAFL." A.A. 311 (May, 1978), 44-47.

100. Duperier, Lieutenant-Colonel. LA VIEILLE EQUIPE. Paris: 1946.

101. Duperier, Bernard. L'ETOILE, LES AILES ET LA COURONNE. Paris: 1954.

102. Ehrengardt, C.J. LES AIGLONS: COMBATS AERIENS DE LA DROLE DE GUERRE. Paris: 1983.

103. Ehrengardt, C.J. and C.F. Shores. L'AVIATION DE VICHY AU COMBAT. Vol. 1. Paris: 1985.

104. Ely, Général Paul. "La leçon qu'il faut tirer des opérations de 1940." R.D.N. (Dec., 1953), 563-582.

105. Facon, Patrick. "Les débuts des forces aériennes françaises libres." A.A. 316 (Dec., 1978), 42-46.

106. ________. "L'armée de l'air d'armistice, de Madagascar au débarquement en Afrique du Nord." M.A. 15 (Dec., 1978), 26-33.

107. ________. "Les forces aériennes françaises libres." A.A. 317 (Jan., 1979), 42-47.

108. ________. "L'armée de l'air française, 1943-1945." A.A. 318 (Feb., 1979), 42-47.

109. ________. "L'armée de l'air française à la veille de la guerre." M.A. 17 (Feb., 1979), 21-30.

110. ________. "L'armée de l'air d'armistice." A.A. 319 (Mar., 1979), 42-45.

111. ________. "L'armée de l'air dans la bataille de 1940." M.A. 19 (Apr., 1979), 11-20.

112. ________. "Réarmement et reconstitution de l'armée de l'air (1942-1943)." R.H.A. (No. 2, 1979), 96-112.

113. ________. "Le plan VII (1943-1944)." R.H.A. (No. 3, 1979), 185-204.

114. ________. "La R.A.F. dans la bataille de france." AVIATION MAGAZINE 774 (Mar. 15, 1980), 70-74; 775 (Apr. 1, 1980), 74-78; 776 (Apr. 15, 1980), 76-78.

115. ________. "Juin 1940: la défaite dans les aires."

AVIATION MAGAZINE 780 (June 15, 1980), 74-77.

116. ________. "L'armée de l'Air reconstitué au combat: le groupe de reconnaissance I/33 Belfort, 1942-1945." M.A. 41 (Feb., 1981), 29-34.

117. ________. "Les forces aériennes françaises en Indochine, 1940-1941." A.A. 342 (July-Aug., 1981), 34-37.

118. ________. "La coopération franco-alliée pendant la campagne de Tunisie. Le problème aérien." R.H.A. (No. 1, 1983), 30-39.

119. ________. "Les forces aériennes françaises libres au Moyen-Orient." CONNAISSANCE DE L'HISTOIRE 53 (Feb., 1983), 6-15.

120. ________. "L'armée de l'Air et le problème du retour en France, 1943-1944." R.H.A. (No. 3, 1984), 63-69.

121. Facon, Patrick and Françoise de Ruffray. "Aperçus sur la collaboration aéronautique franco-allemande, 1940-1943." REVUE D'HISTOIRE DE LA DEUXIEME GUERRE MONDIALE 108 (Oct., 1977), 85-102.

122. Facon, Patrick et Arnaud Teyssier. "Les lecons de la campagne de Pologne vues par l'état-major aérien français: la rapport du Colonel Bergeret, aide-major général des forces aériennes (19 septembre 1939)." R.H.A. (No. 4, 1985), 103-108.

123. FAMILLE AERIENNE. Vichy: 1943.

124. Flamand, Roger. PARIS DE LA FRANCE LIBRE. Paris: 1976.

125. LES FORCES AERIENNES FRANCAISES DE 1939 à 1945, directed by Colonel Pierre Paquier. Paris: 1949.

126. FRERES D'ARMES: HISTOIRE OFFICIELLE DES AVIATIONS DES PAYS OCCUPES. London: 1944.

127. Gallois, P.M. "L'aviation, instrument de l'unification européenne." INTERAVIA 8 (Aug., 1952), 420-424.

128. Gallois, Général Pierre. "Bombardiers lourds, 1943-

1945." R.H.A. (Special number, 1969), 85-98.

129. Geoffre, François de. NORMANDIE-NIEMEN, SOUVENIRS D'UN PILOTE. Paris: 1952.

130. Germain, Louis. MEMOIRES D'UN INCENDIAIRE: SOUVENIRS D'UN BOMBARDIER FRANCAIS DANS LA R.A.F. Paris: 1951.

131. Gisclon, Jean. LE CINQUIEME QUART D'HEURE. Paris: 1965.

132. ________. ILS OUVRIRENT LE BAL. Paris: 1967.

133. ________. LA GRANDE AVENTURE DE LA CHASSE FRANCAISE DE 1939 A 1945. Paris: 1983.

134. Goutard, Colonel A. 1940. LA GUERRE DES OCCASIONS PERDUES. Paris: 1956.

135. Grenier, Fernand. LES AILES FRANCAISES. Algiers: 1944.

136. Grésa, Jacques. COMPLOTS CONTRE L'AVIATION FRANCAISE. UN EPISODE DE LA LUTTE POUR L'INDEPENDANCE NATIONALE. Paris: 1946.

137. Griffin, David. "The Role of the French Air Force: the Battle of France, 1940." A.H. 21 (Fall, 1974), 144-153.

138. Gröhler, Olaf. "Die französische Luftwaffe im Jahre 1940." LUFTVERTEIDIGUNG (No. 5, 1971), 49-53.

139. LE GROUPE DE BOMBARDEMENT TUNISIE DU 8 NOVEMBRE 1942 A LA VICTOIRE. Paris: 1947.

140. Hellade, Jean. LES AILES DANS LA TOURMENTE, 1939-1945. n.p.: 1963.

141. ________. QUAND LE CIEL DE L'AISNE ETAIT EN FEU. Saint-Quentin: 1978.

142. Hayez, Lieutenant-Colonel. "Situation de l'armée de l'air en avions modernes aux premiers jours de la guerre de 1939-1945." F.A.F. 187 (Dec., 1962), 781-794.

143. Heloury, Georges. "Grandeurs et misères de notre aviation en Indochine." L.A. 1048 (Feb. 16, 1946), 3; 1049 (Feb. 23, 1946), 5.

144. HISTORIQUE DE L'ESCADRILLE NORMANDIE-NIEMEN EN URSS: JOURNAL DE MARCHE, 22 MARS 1943-20 JUIN 1945. Paris: 1946.

145. Homolle, Jean. UN PILOTE. Paris: 1947

146. Houart, Georges. EQUIPAGES COURAGEUX. Paris: 1960.

147. Ingold, Gerard. UN MATIN BIEN REMPLI OU LA VIE D'UN PILOTE DE CHASSE DE LA FRANCE LIBRE, 1921-1941. Paris: 1969.

148. Ingrand, I. LE CHEMIN DE CROIX DE L'AVIATION FRANCAISE. Lyon: 1943.

149. Jouhaud, Edmond. LA VIE EST UN COMBAT. SOUVENIRS, 1924-1944. Paris: 1974.

150. Jackson, Robert. AIR WAR OVER FRANCE 1939-1940. London: 1976.

151. Jacobsen, Russell D. "French Airborne Operations during World War II." A.H. 22 (Spring, 1975), 14-22.

152. Jouvenel, Bertrand de. APRES LA DEFAITE. Paris: 1941.

153. Jullian, Marcel. JEAN MARIDOR, CHASSEUR DE V-1. Paris: 1953.

154. Kessel, Joseph. LA BATAILLON DU CIEL. Paris: 1947.

155. Klemm, P.F. "La production aéronautique française de 1940 à 1942." REVUE D'HISTOIRE DE LA DEUXIEME GUERRE MONDIALE 107 (July, 1977), 53-74.

156. Labric, Roger. LA GRANDE ESCADRILLE. Montpellier: 1941.

157. ________. L'ESCADRE INVISIBLE: CEUX DU BOMBARDEMENT DU NUIT. Paris: 1943.

158. ________. ARNOUX, HEROS DES DEUX GUERRES. Paris:

1946.

159. _________. MEMOIRES D'UN AVION DE COMBAT. Paris: 1946.

160. _________. JEAN DAGNAUX, L'AS A LA JAMBE DE BOIS. Paris: 1947.

161. Lacker, Daniel. CARLINGUES SANS GLOIRE. Paris: 1944.

162. Lambermont, P.M. VIDEZ VOS POCHES: HISTOIRE DU GROUPE LORRAINE. Paris: 1954.

163. _________. LORRAINE SQUADRON. Translated by Anthony Pirie. London: 1956.

164. _________. "A propos de l'écran de fumeé du jour J." F.A.F. 108 (Oct., 1955), 615-624.

165. Langeron, Lieutenant-Colonel. MISERE ET GRANDEUR DE NOTRE AVIATION. Paris: 1941.

166. Laubier, Philippe de. "Le bombardement français sur la Meuse, le 14 mai 1940." R.H.A. (No. 3, 1985), 96-109.

167. Lebras, Bertrand. "Les F.A.F.L. (1941-1943)." Unpublished thesis, maitrise, Université de Paris I-Sarbonue, 1980.

168. Leconte, Claude-Henry. ANDRE MOYNET, PILOTE DE COMBAT. Paris: 1955.

169. Le Goaster, Lieutenant-Colonel. "L'action des forces aériennes dans la campagne de France (mai-juin 1940)." REVUE D'HISTOIRE DE LA DEUXIEME GUERRE MONDIALE 10-11 (June, 1953), 135-149.

170. Le Marec, Bernard. LES FRANCAIS LIBRES ET LEURS EMBLEMES. Paris: 1964.

171. Lepotier, Contre-Amiral. "Les avions américains de 1940." F.A.F. 154 (Dec., 1959), 1111-1130.

172. Lesquen, Colonel de. "L'armée de l'air française en 1940." R.D.N. Jan., 1952, 74-84.

173. Levêque, R. "Training French Pilots in the U.S." A.H.

79 (Dec., 1982), 253-259.

174. L'herbier-Montagnon, Germaine. DISPARUS DANS LE CIEL. SOUVENIRS DE LA MISSION DE RECHERCHE DES MORTS ET DES DISPARUS DE L'ARMEE DE L'AIR. Paris: 1943.

175. _________. LA COURONNE T'ATTEND. Paris: 1946.

176. _________. "Le commandant Schloesing." L.A. 1069 (July 13, 1946), 11-12.

177. _________. CAP SANS RETOUR. Paris: 1948.

178. _________. MEMORIAL DES F.A.F.L. DANS L'ACTION ET LE SACRIFICE. Monaco: 1948.

179. "Les lignes aēriennes militaires de la France libre, 1941-1944." ICARE 102 (1982), 7-121.

180. Liron, Jean and Raymond Danel. "The French Air Force, 1939-1940." AIR PICTORIAL 25 (May, 1963), 128-132; (June, 1963), 169-172; (July, 1963), 216-217.

181. Lissy, Marc. CHASSEURS, MES FRERES. Paris: 1947.

182. Livry-Level, Colonel Philippe. MISSIONS DANS LA R.A.F.. Paris: 1946.

183. Livry-Level, Colonel Philippe and Colonel Rēmy. L'OPERATION JERICHO. Paris: 1954.

184. "Avec le Lorraine en Libye." BULLETIN DES FORCES AERIENNES FRANCAISES EN GRANDE-BRETAGNE 2 (1944), 2-7.

185. "Le groupe Lorraine." ICARE 44 (1967-1968), 25-203.

186. Macaigne, Jean. "Il y a deux ans, en juillet 1944, A. de Saint-Exupēry disparaissait... Un des plus riches parmi les hommes." L.A. 1069 (July 13, 1946), 6.

187. MARAUDER FRANCAIS, 1944-1945. Paris: n.d.

188. Mariage, Pierre. LA PASSION DES EQUIPAGES. Paris: 1943.

189. ________. LE DESTIN DES EQUIPAGES. Paris: 1951.

190. ________. "L'adieu aux ailes. La Seconde Guerre mondiale (septembre 1939-10 mai 1940)." F.A.F. 223 (Mar., 1966), 343-356.

191. ________. L'ADIEU AUX AILES. Paris: 1967.

192. Maroselli, André. LE SABOTAGE DE NOTRE AVIATION, CAUSE PRINCIPALE DE NOTRE DEFAITE. Paris: 1941.

193. Martre, Général. "Premiers vols, premiers pas en Italie (mai 1944)." F.A.F. 203 (May, 1964), 627-634.

194. Marzac, M. MAX GUEDJ, LE COMMANDANT AVIATEUR DES F.A.F.L. Casablanca: 1951.

195. Mauléon, Jean. UNE ECOLE DE HEROS. Paris: 1940.

196. Max, Alfred. BLEU R.A.F. Paris: 1951.

Mazare, Marie-Camille. "Indochine 1945: le coup de force japonais et l'armée de l'air." R.H.A. (No. 1, 1978), 123-144.

197. Mendès-France, Pierre. ROISSY-EN-FRANCE. Paris: n.d.

198. Menjaud, Henri. UN GROUPE DE CHASSE AU COMBAT. Grenoble: 1941.

199. Merland, André. A L'ASSAUT DU CIEL: VIE HEROIQUE DU CAPITAINE PIERRE CLAUDE. Paris: 1951.

200. Meyer, Commandant André. AVOIR DES AILES. Nancy: 1967.

201. Michel, Henri. LA DROLE DE GUERRE. Paris: 1971.

202. ________. LE PROCES DE RIOM. Paris: 1979.

203. Migéo, Marcel. LES ROGNEURS D'AILES. Paris: 1941.

204. ________. BATAILLES DANS LE CIEL. Paris: 1943.

205. Mirles, Capitaine. "Normandie." BULLETIN DES FORCES AERIENNES FRANCAISES EN GRANDE-BRETAGNE 1 (1944) 8-11.

206. Mithell, R.H. LES ARMEES DE L'AIR EN PRESENCE. Paris: 1940.

207. Mohrt, Michel. MARIN LE MESLEE. Paris: 1952.

208. Moligny, Dan. LE JOURNAL DE F.L. SMITH-BROWN. Paris: 1946.

209. Mollans, Colonel de. "L'attaque des ponts de la Loire en 1940. Les avions italiens étaient-ils allemands?" MIROIR DE L'HISTOIRE 311 (July-Aug., 1979), 34-43.

210. Mondin, Commissaire. "Le réarmement en Afrique du Nord en 1943." F.A.F. 34 (July, 1949), 34-43.

211. Mouchotte, René. LES CARNETS DE RENE MOUCHOTTE. Paris: 1949.

212. _________. THE MOUCHOTTE DIARIES, 1940-1943. Translated by Philip John Stead. London: 1956.

213. Noetinger, Jacques. CHASSEURS FRANCAIS NES AUX U.S.A. Paris: 1947.

214. Noirot, Général. LES FOUDRES DU CIEL. SOUVENIRS DE GUERRE VECUS DANS LE BOMBER COMMAND PAR UN ANCIEN DE LA ROYAL AIR FORCE. Paris: 1972.

215. "Normandie-Niemen. Tome I: ce que voulait de Gaulle; l'entraînment à Ivanoco; les premiers disparus." ICARE 62 (1972), 34-213.

216. "Normandie-Niemen. Tome II: la bataille d'Orel; les mécaniciens; retour à Toula." ICARE 63 (1972), 34-217.

217. "Normandie-Niemen. Tome III: la campagne de 1944; les mécaniciens soviétiques; la visite du général de Gaulle." ICARE 64 (1973), 30-207.

218. "Normandie-Niemen. Tome IV: la campagne de 1944; la Biélorussie; la bataille du Niemen." ICARE 65 (1973), 30-171.

219. "Normandie-Niemen. Tome V: la campagne de 1945 en Prusse oriental." ICARE 67 (1974), 26-135.

220. "Normandie-Niemen. Tome VI: la victoire; le retour en France." ICARE 70 (1974), 34-149.

221. Paquier, Pierre. AVIATION, MON GRAND AMOUR. Paris: 1944.

222. ________. COMBATS DE CHASSE. Paris: 1946.

223. ________. VISAGES D'AVIATEURS. Paris: 1947.

224. ________. L'AVIATION DE RENSEIGNEMENT FRANCAISE EN 1939-1940. Paris: 1947.

225. ________. L'AVIATION DE BOMBARDEMENT FRANCAISE EN 1939-1940. Paris: 1948.

226. Paquier, Colonel Pierre and Capitaine Joseph-Marie Thonon. LE GROUPE DE BOMBARDEMENT II/12. Paris: 1947.

227. Perret, Louis. EN URSS PENDANT LA GUERRE. LOUIS PERRET, PILOTE AU NORMANDIE-NIEMEN. Paris: 1946.

228. Petit, Général. SOUVENIRS SUR L'ARMEE ROUGE. CONFERENCE PRONONCEE LE 17 JANVIER 1945 AUX JEUDIS DE FRANCE-URSS. Paris: 1945.

229. Piollet, Général. "Renaissance de l'armée de l'air." F.A.F. 15 (Dec., 1947), 300-311.

230. Poincelet, Gilbert. DANS LE CIEL DE SYRIE: L'AVIATION FRANCAISE AU SERVICE DE L'EMPIRE. Paris: 1941.

231. Postel, Claude. "Contre les ponts de la Meuse." R.H.A. (No. 1, 1949), 61-65.

232. ________. "L'aviation française dans le débarquement de Provence." F.A.F. 47 (Aug., 1950), 619-632.

233. Raoul-Duval, C. CIEL DE SABLE. Paris: 1978.

234. Reboul, Colonel. L'ARMEE FRANCAISE DE LA LIBERATION. Paris: 1946.

235. Risso, Colonel. "Le 25e anniversaire de

Normandie-Niemen. Les origines, l'arrivee en URSS, l'entraînement, la première campagne." F.A.F. 246 (Apr., 1968), 533-554.

236. Robert, Chef de Bataillon. "Les parachutistes SAS dans les Côtes du Nord (juin 1944)." R.H.A. (No. 3, 1970), 167-178.

237. Rogé, Lieutenant-Colonel. "Les aviations allemande, française et anglaise du 10 Mai 1940 au 25 juin 1940." R.D.N. Feb., 1951, 162-176.

238. Romans-Petit, Henri. LES MAQUIS DE L'AIN. Paris: 1974.

239. Romeyer, Jean. "Une page d'histoire des forces aériennes françaises libres: l'attaque de Koufra, en Libye, le 5 février 1941." L.A. 1765 (Feb. 20, 1940), 6.

240. Rosez, P. "Le personnel de l'Armée de l'Air (novembre 1942-juillet 1950). Recherche et affirmation d'une identité." Unpublished thesis, Maîtrise, Université de Paris X Nauterre, 1982.

241. Roussel, Philippe. CHASSEURS AU COMBAT. Paris: 1972.

242. Roy, Jules. CIEL ET TERRE. Algiers: n.d.

243. _________. LA VALLEE HEUREUSE. Paris: 1946.

244. _________. THE HAPPY VALLEY. Translated by Edmund Owen Marsh. London: 1952.

245. _________. RETOUR DE L'ENFER. Paris: 1951.

246. _________. RETURN FROM HELL. Translated by Mervyn Savill. London: 1954.

247. _________. PASSION DE SAINT-EXUPERY. Paris: 1951.

248. Saint-Exupéry, Antoine de. PILOTE DE GUERRE. Paris: 1942.

249. _________. FLIGHT TO ARRAS. Translated by Lewis Galantière. London: 1942.

250. ________. ECRITS DE GUERRE, 1939-1944. Paris: 1982.

251. "Saint Exupéry. Quatriène époque: 1939-1940." ICARE 78 (1976), 2-117.

252. "Saint Exupéry. Cinquieme époque: 1941-1943." ICARE 84 (1978), 2-145.

253. "Saint Exupéry. Sixième époque: 1943-1944." ICARE 96 (1981), 1-195.

254. "Saint Exupéry. Toujours vivant." 108 (1984), 8-148.

255. Saint-Geoirs, Etienne. AVIATION FRANCAISE 1942. Lyon: 1942.

256. Saint-Pereuse, Lieutenant-Colonel de. "Le bombardement de la région parisienne le 3 juin 1940." F.A.F. 72 (Sept., 1942), 737-762.

257. Salesse, Lieutenant-Colonel. L'AVIATION DE CHASSE FRANCAISE EN 1939-1940. Paris: 1948.

258. Sauvage, Roger. UN DU NORMANDIE-NIEMEN. Paris: 1959.

259. Segonzac, A. de. "La chasse française libre en action." REVUE DE LA FRANCE LIBRE 210 (Mar.-Apr.-May, 1975), 7-8.

260. Seive, Général Fleury. L'AVIATION D'ASSAUT DANS LA BATAILLE DE 1940. Paris: 1948.

261. Sereau, Raymond. L'ARMEE DE L'ARMISTICE (1940-1942). Paris: 1961.

262. Service Historique de l'Armée de l'Air. "Le groupe Lorraine." F.A.F. 120 (Nov., 1956), 1018-1043.

263. Service Historique de l'Armée de l'Air. "Le groupe de chasse Alsace." F.A.F. 134 (Feb., 1958), 219-237.

264. Severac, E. DANS LE CIEL DE FRANCE. n.p.: 1966.

265. Taillam (?). "Ce que fut l'activité aérienne en AOF pendant la guerre." L.A. 1045 (Jan. 26, 1946), 3.

266. Tessier, Roland. LE BAR DE L'ESCADRILLE. Paris: 1947.

267. ________. LES SACRIFIES DU CIEL. Paris: 1947.

268. ________. CARNETS DE PATROUILLE. Paris: n.d.

269. Teyssier, Arnaud and Charles Christienne. "La reconnaissance de l'Armée de l'Air, 1943-1945." INFOS 117 39 (Dec., 1984), 20-24.

270. Thibaudet, Colonel A. "Réveil de la chasse française, 1942-1945." F.A.F. 50 (Nov., 1950), 196-223.

271. Tillon, Charles. RENAISSANCE DE L'AVIATION FRANCAISE. Paris: 1944.

272. Truchet, A. "L'armistice de juin 1940 et l'Afrique du Nord." REVUE D'HISTOIRE DE LA DEUXIEME GUERRE MONDIALE 3 (June, 1951), 27-50.

273. Valin, Général Martial. "Maurice Max Guedj." F.A.F. 76 (Jan., 1953), 489-494.

274. ________. LES SANS-CULOTTES DE L'AIR. Paris: 1954.

275. Verrier, Marcel. ARENES DU CIEL. Paris: 1967.

276. Vibraye, Tony de. AVEC MON GROUPE DE RECONNAISSANCE: AOUT 1939-AOUT 1940. Roanne: 1943.

277. Vigneras, Marcel. REARMING THE FRENCH. Washington: 1957.

278. Vilain, C. L'ESCADRILLE DES SIOUX. Paris: 1940.

279. Weil, Gérard. "Max Guedj." F.A.F. 263 (Nov., 1969), 391-418.

280. Williame, Capitaine. L'ESCADRILLE DES CIGOGNES, SPA 3, 1939-1940. Paris: 1945.

281. Williams, Major. AIR POWER. New York: 1940.

282. Yonnet, Henri and Yves Laty. LE JULES VERNE, AVION CORSAIRE. Paris: n.d.

283. Zorn, J. L'EPOPEE DE L'ESCADRILLE NORMANDIE-NIEMEN. Paris: 1946.

284. __________. AVEC CEUX DU GROUPE BRETAGNE. Paris: 1947.

THE PERIOD SINCE 1945

The essential works of a general nature relating to this period are those of Claude Carlier, his theses (105, 107) and his published history L'AERONAUTIQUE FRANCAISE 1945-1975 (108). These works by their very nature offer only a broad survey of military and civil aviation. There is also a very useful special number of the REVUE HISTORIQUE DES ARMEES covering the Armée de l'Air from 1945 to 1949, prepared by the Service Historique and published in 1982. The various articles in that number are listed individually in the bibliography which follows, and they cover such diverse subjects as organization, doctrine, the aeronautical industry, the budget, schools, the various combat specialties, and the use of the air forces in Indochina. These articles can be completed by that of J.P. Dournel on the situation in 1946 (169) and that of Patrick Facon on the Armée de l'Air and the Dunkirk and Brussels treaties, (204). Finally, the general evolution of the Armée de l'Air may be followed in several periodicals: LES AILES, especially for the late forties and fifties, the REVUE DE DEFENSE NATIONALE, and FORCES AERIENNES FRANCAISES until the end of the sixties.

The transformations in the aircraft industry are treated in the works of Claude Carlier, already mentioned, the article by Emmanuel Chadeau on industrial problems in the late forties (114), the report of Inspector General Pellenc (370), the paper of General Gallois on the five year plan of 1951-1955 (233) and the other published papers of the 1985 colloquium DE L'AERONAUTIQUE A L'ESPACE: 40 ANNEES DE DEVELOPPEMENT AEROSPATIAL FRANCAIS, 1945-1985 (these papers are listed individually in this bibliography).

The bibliography on the Indochina War is relatively sparse. Véronique Debernardy is preparing a thesis on the adaptation of the Armée de l'Air to anti-guerilla operations in the Far East from 1945 to 1954; she has already completed a manuscript study of the Battle of Na San which is in the S.H.A.A. collections. Among general books on the Indochina War General Chassin's AVIATION INDOCHINE (127) is at once technical and anecdotal in treatment. Certain aspects of the air effort have been treated by Patrick Facon, and his series of studies in LE MONITEUR DE L'AERONAUTIQUE (189-197) can be consulted with profit; finally, various issues of FORCES AERIENNES FRANCAISES are useful, especially two special numbers on the Indochina War which appeared in 1952 and 1953.

The Algerian War has as yet been little treated in scientific works, since the bulk of the archives on this conflict are still closed. There is relatively little to be called to the reader's attention, save a few detailed studies, such as those of Patrick Facon on light aviation and helicopters in Algerian operations (198, 199).

The Suez expedition has not yet been the object of important studies relating to the Armée de l'Air, but those who wish to explore the subject would do well to begin with the work of General Beaufre (27) and of M. and S. Bronberger (96). There was also a special number of the REVUE HISTORIQUE DES ARMEES dedicated to the expedition (No. 4, 1986).

As for the problems of nuclear armament as they relate to the Armée de l'Air, these can be seen in works by General Gallois and C. Delmas (159, 230). An important colloquium was held in Besanson in September, 1984. Its published proceedings reveal much about General de Gaulle's ideas on dissuasion and on the various stages in the development of France's nuclear forces. General Robineau has authored a useful study of the adaptation of the Armée de l'Air to the nuclear era (396); finally the subject is treated in many articles in FORCES AERIENNES FRANCAISES.

BIBLIOGRAPHICAL LISTING

1. "L'action aérienne et les opérations terrestres." R.D.N. Oct., 1950, 275-285; Nov., 1950, 414-422.

2. "Les activités et les missions du GMMTA au cours de 1957." L.A. 1670 (Mar. 1, 1955), 12-13.

3. "Algeria First Stalls French Air Pace." AVIATION WEEKLY. Mar. 9, 1959, 275-276.

4. André, Valérie. "Les hélicoptères en Indochine." F.A.F. 87 (supplement), 611-615.

5. _________. ICI VENTILATEUR. Paris: 1954.

6. "L'armée de l'air en action: opérations en Indochine et à Madagascar." L.A. 1179 (Sept. 4, 1948), 11.

7. "L'armée de l'air à l'heure de l'austérité." L.A. 1665 (Jan. 25, 1958), 11-12.

8. "L'armée de l'air dans les opérations de pacification en Algérie." L.A. 1619 (Feb. 9, 1957), 11.

9. Assouline, Pierre. MONSIEUR DASSAULT. Paris: 1983.

10. "Avec l'armée de l'air en Indochine: de la liquidation japonaise à nos jours." L.A. 1495 (Sept. 25, 1954), 6-7.

11. "Avec l'armée de l'air en Indochine: regards vers l' arrière, 1945-1946." L.A. 1494 (Sept. 17, 1954), 6.

12. Arpurt, Lieutenant-Colonel. "Aspects des opérations héliportées en Algérie." F.A.F. 158 (Apr., 1960), 637-662.

13. "L'assemblée nationale devant les crédits de l'armée de l'air." L.A. 1852 (Nov. 17, 1961), 7.

14. "L'aviation sur le canal de Suez: la maîtrise de l' air, condition du débarquement." L.A. 1606 (Nov. 10, 1956), 7.

15. "Des avions pour l'Indochine." L.A. 1213 (Apr. 30, 1949), 5.

16. Azeau, Henri. LE PIEGE DE SUEZ. Paris: 1964.

17. Bachelier, Lieutenant-Colonel. "L'armée de l'air et le centre d'expérimentations du Pacifique." F.A.F. 233 (Feb., 1967), 161-182.

18. Backlund, Donald R. "Stalingrad and Dien Bien Phu: Two Cases of Failure in Strategic Resupply." A.H. 17 (Summer, 1970), 60-68.

19. Bail, René. HELICOPTERES ET COMMANDOS DE MARINE EN ALGERIE. Paris: 1983.

20. Baille, Claude. LES EQUIPAGES SANS NOM. Châteaudun: 1968.

21. Barthélémy, Lieutenant-Colonel. "Politique du personnel." F.A.F. 35 (Aug., 1949), 641-659.

22. Barthélémy, Général. "Il faut rénover notre transport aérien militaire." F.A.F. 175 (Nov., 1961), 627-638.

23. _________. "Le COTAM." F.A.F. 180 (Apr., 1962), 539-548.

24. _________. HISTOIRE DU TRANSPORT AERIEN MILITAIRE FRANCAIS. Paris: 1981.

25. _________. "L'aviation de transport militaire française de 1945 à 1949." R.H.A. special number, 1982, 70-81.

26. Basset, Lieutenant-Colonel. "La formation des officiers d'état major de l'armée de l'air." F.A.F. 7 (Apr., 1947), 53-63.

27. Beaufre, Général. L'EXPEDITION DE SUEZ. Paris: 1967.

28. Benavente, J.P. "Aux premiers temps des hélicoptères." A.A. 316 (Dec., 1978), 36-41.

29. Blanc, Edmond. "Le général Valin." L.A. 1224 (July 16, 1949), 5.

30. ________. "Manoeuvres aériennes en Afrique noire." L.A. 1103 (Mar. 13, 1947), 4.

31. ________. "La nouvelle mission de l'armée de l'air." L.A. 1120 (July 12, 1947), 4; 1121 (July 19, 1947), 4-5.

32. ________. "Renaissance des écoles de l'air." L.A. 1139 (Nov. 23, 1947), 4.

33. ________. "Avec le général Léchères à travers l'Union française." L.A. 1144 (Jan. 3, 1948), 1-5.

34. ________. "L'oeuvre d'une escadrille sur la Côte des Somalis." L.A. 1203 (Feb. 19, 1949), 7.

35. ________. "Quand le Général Hartemann parle des programmes de l'armée de l'air." L.A. 1256 (Feb. 25, 1950), 4-5.

36. ________. "Pour défendre la France, il lui faut une puissante armée de l'air." L.A. 1261 (Apr. 1, 1950), 5-6.

37. ________. "Créer des usines en Afrique?" L.A. 1263 (Apr. 15, 1950), 6.

38. BLANC (Michel). "Recherche sociologique sur la mobilité sociale et professionnelle des sous-officiers de l'armée de l'air: essai d' application d'un modèle d'analyse de type stratégique." Unpublished thesis, doctorat de IIIe cycle in sociology, Université de Paris V, 1974.

39. Bodet, Général. "La dissolution de la 11e brigade de bombardement moyen." F.A.F. 115 (May, 1956), 811-838.

40. Bodin, Michael. "Le corps expéditionnaire français en Indochine." Unpublished thesis, doctorat d'état, Université de Paris I, 1979.

41. Boiville, Sous-Lieutenant de. "La SPA-3 en Indochine." F.A.F. 13 (Oct., 1947), 47-50.

42. Borand, Charles-André. "Le point bas de notre aviation." L.A. 1247 (Dec. 24, 1947), 6.

43. ________. "Le meilleur atout de la défense." L.A. 1228 (Aug. 13, 1948), 4.

44. ________. "La force d'arrêt. Sa mission essentielle et capitale sera de défendre le ciel français jusqu'à l'arrivée des forces alliées." L.A. 1186 (Oct. 23, 1948), 6.

45. ________. "La constitution d'une infrastructure sera, pour la France, une opération rentable et notre pays deviendra ainsi un chaînon dans la chaîne de défense occidentale." L.A. 1189 (Nov. 13, 1948), 5.

46. ________. "Veut-on tuer notre aviation?" L.A. 1192 (Dec. 4, 1948), 6.

47. ________. "L'organisation de notre armée de l'air au sein de la coalition alliée." L.A. 1193 (Dec. 11, 1948), 6.

48. ________. "Faire confiance à notre armée de l'air." L.A. 1196 (Jan. 1, 1949), 6-7.

49. ________. "La formation du personnel." L.A. 1198 (Jan. 15, 1949), 7.

50. ________. "Le défaitisme du ciel." L.A. 1200 (Jan. 29, 1949), 5.

51. ________. "Une armée de cadres." L.A. 1201 (Feb. 5, 1949), 5.

52. ________. "Les données particulières de notre défense aérienne." L.A. 1202 (Feb. 12, 1949), 5.

53. ________. "Le commandement doit être aérien." L.A. 1206 (Mar. 12, 1949), 6-7.

54. ________. "Le grand danger." L.A. 1207 (Mar. 19, 1949), 5.

55. ________. "Le bombardier est nécessaire." L.A. 1208 (Mar. 26, 1949), 5.

56. ________. "Défense nationale et forces navales." L.A. 1209 (Apr. 2, 1949), 6.

57. ________. "L'aviation de la France." L.A. 1213 (Apr. 30, 1949), 7.

58. ________. "Le salon de l'espoir." L.A. 1215 (May 14, 1949), 7.

59. ________. "Le programme doit être complet." L.A. 1216 (May 21, 1949), 7.

60. ________. "Et le vote du budget?" L.A. 1217 (May 28, 1949), 5.

61. ________. "Les vraies leçons du salon." L.A. 1219 (June 11, 1949), 5.

62. ________. "Débats touffus et creux." L.A. 1222 (July 2, 1949), 5.

63. ________. "Le temps est venu de réaliser." L.A. 1224 (July 16, 1949), 6.

64. ________. "Avions d'armes de demain." L.A. 1224 (July 16, 1949), 5-6.

65. ________. "Le Congo, base de notre redressement." L.A. 1225 (July 23, 1949), 6.

66. ________. "Les armes auxiliaires." L.A. 1226 (July 30, 1949), 11.

67. ________. "Un plan doit être complet." L.A. 1227 (Aug. 6, 1949), 5.

68. ________. "Appel au parlement." L.A. 1229 (Aug. 20, 1949), 4.

69. ________. "La stratégie des comptables." L.A. 1240 (Aug. 27, 1949), 4-5.

70. ________. "Le mauvais rêve." L.A. 1231 (Sept. 3, 1949), 5-6.

71. ________. "La relève nécessaire." L.A. 1232 (Sept. 10, 1949), 5.

72. ________. "Engin et avion." L.A. 1234 (Sept. 24, 1949), 5.

73. ________. "La conduite de la guerre." L.A. 1235 (Oct. 1, 1949), 5.

74. ________. "La France dans la coalition." L.A. 1240 (Nov. 5, 1949), 6.

75. ________. "Le but: éviter l'invasion." L.A. 1241 (Nov. 12, 1949), 5.

76. ________. "La faculté d'imaginer." L.A. 1242 (Nov. 19, 1949), 6-7.

77. ________. "Avant que soit voté le budget." L.A. 1244 (Dec. 3, 1949), 6.

78. ________. "Un plan, une loi." L.A. 1245 (Dec. 10, 1949), 6.

79. ________. "Les principes de base." L.A. 1246 (Dec. 17, 1949), 6.

80. ________. "Un tout: l'armée-nation." L.A. 1249 (Jan. 7, 1950), 6.

81. ________. "Un jugement clairvoyant." L.A. 1251 (21 Jan., 1950), 6-7.

82. ________. "La France n'a pas su choisir." L.A. 1253 (Feb. 4, 1950), 6-7.

83. ________. "Une armée de l'air doit être complète." L.A. 1254 (Nov. 11, 1950), 6.

84. ________. "Le fond véritable du problème." L.A. 1255 (Feb. 18, 1950), 5-6.

85. ________. "Défense aérienne dans l'Union française." L.A. 1256 (Feb. 25, 1950), 5-6.

86. ________. "L'organisation de notre défense." L.A. 1260 (Mar. 25, 1950), 6-7.

128. "Le choix de notre bombardier léger." L.A. 1287 (Sept. 30, 1950), 1.

129. Chombart, de Lauwe P. "L'école de l'air en 1946." F.A.F. 10 (July, 1047), 492-504.

130. Christiaens, L. "Il faut refaire l'unité de l'aviation française." L.A. 1523 (Apr. 9, 1955), 2.

131. Christiaens, Général Charles. "L'armée de l'air de 1945 à 1949: le contexte général." R.H.A. 1982, special number, 4-15.

132. ________. "Un chef d'état-major général de l'armée de l'air: le général Gérardot." R.H.A. 1982 Special number, 96-101.

133. ________. "L'armée de l'air en mai 1945." in DE GAULLE ET LA NATION FACE AUX PROBLEMES DE DEFENSE; 1945-1946. Paris: 1983, 43-46.

134. Christienne, Général Charles and Patrice Buffotot. "L' adaptation de l'armée de l'air française à un type de guerre nouveau: la guerre d'Indochine (1946-1954)." REVUE INTERNATIONALE D'HISTOIRE MILITAIRE 49 (1980), 183-190.

135. Claude, Jacques. "L'opération combinée air-terre." F.A.F. 159 (May, 1960), 781-788.

136. Clerc, Jean. "Avec le groupe Béarn en Indochine." L.A. 1117 (June 21, 1947), 12.

137. Clostermann, Pierre. APPUI FEU SUR L'OUED HALLAIL. Paris: 1960.

138. "Le commandement des écoles de l'armée de l'air." A.A. 300 (Mar., 1976), 34-42.

139. Condom, Pierre. "L'armée de l'air: un élément essentiel de notre politique nationale." AVIATION MAGAZINE 606 (Mar. 15, 1973), 22-47.

140. Copel, Général E. VAINCRE LA GUERRE. Paris: 1983.

141. Couilleau, Commandant. "Un an de transport aérien en Indochine." F.A.F. 122 (Jan., 1957), 113-126.

87. ________. "Les transports aériens dans l'Union française." L.A. 1258 (Mar. 11, 1950), 6.

88. Breuillac, Général. "La défense de la France." R.D.N. June, 1948, 777-798.

89. Briand, G. "Il faut instruire les réservistes aux fonctions nouvelles qu'on leur assigne." L.A. 1155 (Mar. 20, 1948), 11.

90. Bronberger, Merry and Serge. LES SECRETS DE L' EXPEDITION D'EGYPTE. Paris: 1957.

91. Brossard, Commandant. LES AILES DE NEPTUNE. Paris: 1959.

92. Bousquet, R. "La force nucléaire stratégique française." R.D.N. May, 1966, 793-811.

93. Bousquet, Ingénieur-Général. "La direction technique des constructions aéronautiques." R.D.N. June, 1979, 41-50.

94. Buchalet, Commandant. "L'arme aéroportée et la réorganisation de l'armée française." R.D.N. May, 1946, 639-653.

95. Cabrière, J. "La géomètrie variable." R.D.N. June, 1969, 994-1007.

96. Callet, Lieutenant-Colonel Jean. HIVER A TEBESSA. Paris: 1959.

97. Calmel, Colonel J. "Aviation de coopération." R.D.N. Aug.-Sept., 1956, 1085-1096.

98. ________. "Bilan 1956." F.A.F. 121 (Dec., 1956), 1254-1283.

99. ________. "Politique à long terme des matériels aériens." F.A.F. 124 (Mar., 1957), 545-553.

100. ________. "Renaissance du bombardement." F.A.F. 127 (June, 1957), 1097-1110.

101. ________. "Nouvelle politique aérienne de la France." F.A.F. 128 (July, 1957), 147-152.

102. Calmel, Général J. "Présence de l'armée de l'air en Algérie en 1958." F.A.F. 146 (Mar., 1959), 439-444.

103. ________. "Emploi de l'armée de l'air en Algérie." F.A.F. 149 (June, 1959), 1060-1062.

104. ________. "Les forces aériennes tactiques." F.A.F. 213 (Apr., 1956), 441-454.

105. Carlier, Claude. "Le renouveau de l'aéronautique militaire en France sous la Quatrième République (1946-1958)." Unpublished thesis, maitrîse, Université Paul Valéry-Montpellier III, 1972.

106. ________. LE DEVELOPPEMENT DE L'AERONAUTIQUE MILITAIRE FRANCAISE DE 1958 A 1970. Vincennes: 1979.

107. ________. "L'aéronautique militaire française (1945-1975)." Unpublished thesis, doctorat d'état, Université Paul Valéry-Montpellier III, 1978.

108. ________. L'AERONAUTIQUE FRANCAISE: 1945-1975. Paris: 1983.

109. Cartier, Paul. "Revenir au ministère de l'air." L.A. 1194 (Dec. 18, 1948), 6.

110. ________. "L'aviation militaire française en danger." L.A. 1237 (Oct. 15, 1949), 6.

111. Castaing, Commissaire-Général. "Le commissariat de l'air et la vie économique en AOF." F.A.F. 118 (Aug.-Sept., 1956), 377-392.

112. Castex, Louis. "Le sort du plan quinquennal." L.A. 1262 (Apr. 8, 1950), 6-7.

113. Cauchetier, Lieutenant Raymond. "Na San." F.A.F. 87, supplement, 569-580.

114. Chadeau, Emmanuel. "Volume et emploi des dépenses aéronautiques 1945-1950." R.H.A. (no. 1, 1982), 24-41.

115. ________. "Notes sur les problèmes industriels aéronautique nationale, 1944-1949." R.H.A. Special number, 40-49.

116. Chalandon (Ingénieur en chef de l'air). "Intercep légers et avions expérimentaux français à ai delta." F.A.F. 125 (Apr., 1957), 757-767.

117. ________. "L'industrie aéronautique française su voie de l'expansion et de la coopération européenne." F.A.F. 127 (June, 1957), 1001-1

118. ________. "Bilan d'une année dans l'industrie aéronautique française." F.A.F. 137 (May, 19 661-696.

119. ________. "Bilan et perspectives de l'industrie aéronautique française." F.A.F. 160 (June, 19 937-964.

120. ________. "L'industrie aéronautique française: aperçus sur la vie des principales sociétés aéronautiques." F.A.F. 173 (Aug.-Sept., 1961) 231-276.

121. ________. "Progrès et évolution de l'industrie aéronautique française." F.A.F. 182 (June, 19 861-886.

122. "Des chars ou des avions?" R.D.N. July, 1950, 3-11.

123. Charriou, André. "Quinze ans d'industrie aéronauti française." R.D.N. Dec., 1948, 649-666.

124. Chassagny, Marcel. "Politique industrielle de l'aéronautique." R.D.N. June, 1969, 973-979.

125. Chassin, Général L.M. "Questions essentielles." F.A 6 (Mar., 1947), 735-748.

126. ________. "Un an d'opérations aériennes (avril 1952-avril 1953)." F.A.F. 87, supplement, 487-504.

127. ________. AVIATION INDOCHINE. Paris: 1954.

142. Coulet, Commandant. "Les commandos de l'air." F.A.F. 122 (Jan., 1957), 127-138.

143. Coupet, Lucien. "Polémique nocive." L.A. 1223 (July 9, 1947), 4-5.

144. "Les crédits de l'armée de l'air devant l'assemblée nationale constituante." L.A. 1056 (Apr. 13, 1946), 1-3.

145. Crépin, Général. "Pour une aviation d'appui." R.D.N. Aug.-Sept., 1953, 159-165.

146. ________. "A propos d'une aviation légère d'appui." F.A.F. 92 (May, 1954), 600-676.

147. ________. "Armement, stratégie et tactique." R.D.N. June, 1969, 980-993.

148. Cros, Général. "Le génie de l'air en Algérie et au Sahara." R.H.A. (No. 4, 1959), 156-158.

149. ________. "Le génie de l'air, 1950-1960." F.A.F. 160 (June, 1960), 917-936.

150. Darde, Lieutenant-Colonel. "Le statut des officiers du personnel navigant." F.A.F. 131 (Nov., 1957), 567-578.

151. Dassault, Marcel. "Le point de notre aviation." L.A. 1245 (Dec. 10, 1949), 4.

152. ________. THE TALISMAN: THE AUTOBIOGRAPHY OF MARCEL DASSAULT, CREATOR OF THE MIRAGE JET. New Rochelle: 1971.

153. Davet, René. "Ceux qui firent notre armée de l'air: le général Paul Gérardot." L.A. 1313 (Mar. 31, 1951), 5-6.

154. Debernardy, Véronique. "Les ecoles de l'armée de l'air de 1945 à 1949." R.H.A. 1982, special number, 50-59.

155. Debrabant, Lieutenant-Colonel. "Nécessité et possibilité d'une force de bombardement française." F.A.F. 48 (Sept., 1950), 719-746.

156. Delachoue, Lieutenant. "Hélicoptères en Indochine." F.A.F. 67 (Apr., 1952), 64-82.

157. Delaye, Colonel. "L'aviation et notre centre nucléaire du Pacifique." F.A.F. 222 (Feb., 1966), 171-184.

158. Delmas, Claude. "La force de frappe nationale: les données du problème." R.D.N. Oct., 1960, 1549-1565.

159. ________. HISTOIRE POLITIQUE DE LA BOMBE ATOMIQUE. Paris: 1967.

160. ________. ARMEMENTS NUCLEAIRES ET GUERRE FROIDE. Paris: 1971.

161. Delmer, Christophe. "Coup d'oeil sur les SSBS." F.A.F. 223 (Mar., 1966), 357-362.

162. Deleuze, Commandant and Commandant Héliot. "Invaders en Indochine." F.A.F. 87, supplement, 553-568.

163. Demoulin, Pierre. "Le bombardier supersonique Mirage IVA." L.A. 1766 (Feb. 27, 1960), 8, 9.

164. Desbordes, Pierre. "Le Nord 2501." L.A. 1305 (Feb. 3, 1951), 8-9.

165. Desclaire, Jean. "Le corps du génie de l'air." R.D.N. Jan., 1947, 60-74.

166. Devillers, Colonel. "Hélicoptères en Algérie." F.A.F. 122 (Jan., 1957), 80-92.

167. Devriese, Jean. "Le moteur d'avion français a-t-il sa chance dans le monde?" R.D.N. June, 1969, 1008-1018.

168. Dollfus, Daniel. LA FORCE DE FRAPPE. Paris: 1960.

169. Dournel, Jean-Pierre. "L'armée de l'air en 1946." REVUE D'HISTOIRE DE LA DEUXIEME GUERRE MONDIALE. 110 (Apr., 1978), 25-44.

170. Dupouy, Commandant. "L'aviation vietnamienne." F.A.F. 87, supplement, 581-595.

171. Duval, Colonel R. "Peut-on améliorer l'efficacité opérationnelle de l'armée de l'air?" F.A.F. 125 (Apr., 1957), 628-664.

172. "Les écoles de l'armée de l'air et la formation du personnel spécialisé." L.A. 1661 (Dec. 28, 1957), 13-14.

173. "L'école de formation initiale du personnel navigant." A.A. 295 (Mar., 1975), 55-57.

174. "L'Ecole des navigateurs." F.A.F. 283 (Aug.-Sept., 1971), 153-160.

175. "L'école technique de l'armée de l'air." A.A. 295 (Mar., 1975), 44-49.

176. "L'école de Saintes." A.A. 301 (May, 1976), 32-39.

177. "L'effort de l'armée de l'air en Algérie au point de vue infrastructure." L.A. 1625 (Mar. 23, 1957), 12-13.

178. Ely, Général. MEMOIRES. Paris: 1969.

179 "Embargo Dims French Aerospace Outlook." AVIATION WEEK AND SPACE TECHNOLOGY. Jan. 27, 1969, 21-22.

180. "Entretien avec le général C.L. Grigaut, chef d'état-major de l'air." R.D.N. Oct., 1974, 75-92.

181. Escabasse, Commandant. "Développement du transport aérien en Afrique française." F.A.F. 90 (Mar., 1954), 345-362.

182. ________. "Contre une aviation légère d'appui." F.A.F. 88 (Jan., 1954), 65-74.

183. ________. "Vocation à la stratégie aérienne des territoires français d'Afrique." F.A.F. 106 (July, 1955), 137-163.

184. Escurat, Lieutenant-Colonel. "Le bureau central d'incorporation et d'archives de l'armée de l'air." A.A. 295 (Mar., 1975), 42-43.

185. Fabry, Général. "Le 2e CATAC." F.A.F. 213 (Apr., 1965), 479-484.

186. Facon, Patrick. "1946. La reconstruction de l'armée de l'air." M.A. 28 (Jan., 1980), 27-33.

187. ________. "1946. Le problème industriel: des avions pour l'armée de l'air." M.A. 29 (Feb., 1980), 11-18.

188. ________. "L'armée de l'air à l'heure des choix, 1946-1948." M.A. 30 (Mar., 1980), 26-35.

189. ________. "L'armée de l'air en Indochine: la reconstitution, mai 1945-mars 1946." M.A. 31 (Apr., 1980), 11-17.

190. ________. "L'armée de l'air en Indochine: la pacification, mars 1946-mai 1947." M.A. 32 (May, 1980), 40-44.

191. ________. "L'armée de l'air en Indochine: L'impasse, mai 1947-octobre 1949." M.A. 34 (July, 1980), 19-24.

192. ________. "L'armée de l'air en Indochine: L' enlisement, janvier 1950-janvier 1952." M.A. 35 (Aug., 1980), 28-34.

193. ________. "L'armée de l'air en Indochine: l'éviction, février 1952-juillet 1954." M.A. 36 (Sept., 1980), 44-50.

194. ________. "L'armée de l'air en Indochine: les avions de chasse, 1945-1954." M.A. 37 (Oct., 1980), 21-27.

195. ________. "L'armée de l'air en Indochine: les avions de bombardement, 1945-1954." M.A. 38 (Nov., 1980), 14-21.

196. ________. "L'armée de l'air en Indochine: les avions de transport, de reconnaissance et d'observation, 1945-1954." M.A. 39 (Dec., 1980), 25-31.

197. ________. "L'armée de l'air en Indochine: les hélicoptères, 1950-1954." M.A. 40 (Jan., 1981), 10-16.

198. ________. "L'aviation légère dans la guerre d'Algérie, 1954-1962." M.A. 47 (Aug., 1981), 21-31.

199. ________. "Les hélicoptères de l'Armée de l'air en Algérie" CONNAISSANCE DE L'HISTOIRE 46 (June, 1982), 6-15.

200. ________. "Reconstitution de l'Armée de l'air en Indochine 1945-1946." R.H.A. 1982, Special number, 82-89.

201. ________. "L'Armée de l'air et le traité de Bruxelles, 1947-1948." R.H.A. 1982, Special number, 90-95.

202. ________. "Dien Bien Phu dans les airs." CONNAISSANCE DE L'HISTOIRE 58 (July, 1983), 4-19.

203. ________. "L'Armée de l'air, sa place dans la défense nationale." in DE GAULLE ET LA NATION FACE AUX PROBLEMES DE DEFENSE; 1945-1946. Paris: 1983, 205-212.

204. ________. "L'Armée de l'air française à l'heure des pactes de Dunkerque et de Bruxelles, 1947-1948." COLLOQUE ARMEES ET SYSTEMES D'ALLIANCE. Montpellier: 1984, 969-989.

205. ________. "Dien Bien Phu." AVIATION MAGAZINE 875 (1984), 72-75; 876 (1984), 74-78.

206. ________. "L'Armée de l'air et Dien Bien Phu." R.H.A. (No. 4, 1984), 58-64; (No. 1, 1985), 79-87.

207. Facon, Patrick and Arnaud Teyssier. CINQUANTE ANS DE L'ECOLE DE L'AIR. Paris: 1985.

208. Fatio, Capitaine Pierre. "Formation des pilotes vietnamiens." F.A.F. 110 (Dec., 1955), 1025-1049.

209. Faucanie, Commandant. "Organisation du commandement de l'air en AOF: étude d'un déploiement adapté à la mission." F.A.F. 118 (Aug.-Sept., 1956), 319-344.

210. Fay, Général. "Les ailes françaises en Indochine." F.A.F. 7 (Apr., 1947), 5-24.

211. Flament, M. LES HELICOS DU DJEBEL. ALGERIE, 1955-1962. Paris: 1982.

212. Fleurquin, Général P.A. "Importance stratégique des bases aériennes de l'AOrF et ses conséquences." F.A.F. 106 (July, 1955), 83-88.

213. ________. "Evolution du rôle des commandements de l'air dans les territoires d'outre-mer." F.A.F. 117 (July, 1956), 51-66.

214. "Les forces aériennes stratégiques." A.A. 298 (Nov., 1975), 14-19.

215. "Force de frappe et bouclier." L.A. 1751 (Nov. 14, 1959), 7.

216. "Force de Frappe. A Triple Debate." ATLANTIC COMMUNITY QUARTERLY 2 (1964), 43-46.

217. Forestier, Laurent. "La bataille de Dien Bien Phu." A.A. 308 (Feb., 1978), 41-47; 309 (Mar., 1978), 44-48.

218. "La formation des équipages de l'aviation légère." L.A. 1656 (Nov. 23, 1957), 11-12.

219. Fouquières, Lieutenant-Colonel L. de. "L'effort de l'armée de l'air en 1952." F.A.F. 76 (Jan., 1953), 580-588.

220. ________. "Un an de lutte en Indochine." F.A.F. 87, supplement, 603-610.

221. ________. "L'armée de l'air, bilan 1954." F.A.F. 99 (Dec., 1954), 1074-1083.

222. ________. "Les hélicoptères en Indochine." F.A.F. 100 (Jan., 1955), 227-231.

223. ________. "Union française et armée de l'air." F.A.F. 106 (July, 1955), 195-203.

224. Frachet, André. "Le SO-6020, chasseur à réaction." L.A. 1220 (June 18, 1946), 8.

225. _________. "Le bombardier Bréguet 482." L.A. 1155 (Mar. 20, 1948), 8.

226. _________. "L'avion-cargo Nord 2500." L.A. 1221 (June 25, 1949), 8.

227. _________. "Le chasseur Nord 2200." L.A. 1249 (Jan. 7, 1950), 8.

228. Fredet, Commandant. "Le problème des écoles de formation du personnel navigant." F.A.A. 35 (Aug., 1949), 613-640.

229. "The French Air Force." FLIGHT 67 (May 13, 1955), 625-627.

230. Gallois, Général Pierre. LA GRANDE BERNE: L'ATOME ET LES NEGOCIATIONS EST-OUEST. Paris: 1975.

231. _________. L'ADIEU AUX ARMEES. Paris: 1976.

232. _________. LE RENONCEMENT DE LA FRANCE DEFENDUE A L'EUROPE PROTEGEE. Paris: 1977.

233. _________. "Le plan quinquenal aéronautique (1951-1955) et ses conséquences." COLLOQUE DE L'AERONAUTIQUE A L'ESPACE. 40 ANNEES DE DEVELOPPEMENT AEROSPATIAL FRANCAIS: 1945-1985. Paris: 1986, 29-41.

234. Gatel, Lieutenant-Colonel. "Forces aériennes françaises en Afrique occidentale." F.A.F. 150 (July, 1959), 115-130.

235. Génévey, ingénieur général de l'air Louis. "L'industrie aérospatiale et la planification." F.A.F. 204 (June, 1964), 731-804.

236. Gensé, Commandant M. "L'armée de l'air à l'âge de l'engin." F.A.F. 115 (May, 1956), 785-810.

237. Genty, Commandant. "Le transport aérien d'assaut dans le cadre des forces françaises." F.A.F. 81 (June, 1953), 295-310.

238. ________. "L'industrie aéronautique en France." F.A.F. 100 (Jan., 1955), 55-91.

239. Gérardot, Général Paul. "La bataille aérienne." R.D.N. May, 1948, 584-596.

240. ________. "La défense de l'Europe occidentale." R.D.N. Oct., 1949, 293-307.

241. ________. "Avant tout, gagner la bataille aérienne." L.A. 1264 (Apr. 22, 1950), 1, 6.

242. ________. "Quelles sont les possibilités actuelles de l'industrie aéronautique française?" L.A. 1295 (Nov. 25, 1950), 7.

243. ________. "Constitution d'une armée de l'air française." R.D.N. Dec., 1950, 543-557.

244. ________. "Production aéronautique française et livraisons en provenance de l'étranger." L.A. 1296 (Dec. 2, 1950), 6.

245. ________. "Règles essentielles d'une politique des prototypes." L.A. 1298, (Dec. 16, 1950), 6.

246. ________. "L'aviation légère au service de la défense." L.A. 1302 (Jan. 13, 1951), 6.

247. ________. "Il faut laisser s'épanouir la pensée militaire française." L.A. 1305 (Feb. 3, 1951), 6.

248. ________. "La valeur de notre armée de l'air dépend de sa valeur intellectuelle." L.A. 1306 (Feb. 10, 1951), 6.

249. ________. "Le recrutement du personnel navigant." L.A. 1334 (Aug. 25, 1951), 6.

250. ________. "Le problème angoissant du recrutement: une solution, l'inscription aérienne." L.A. 1343 (Oct. 27, 1951), 6.

251. ________. "Cinq cents avions légers en Indochine." L.A. 1346 (Nov. 13, 1951), 6.

252. ________. "Pour une nouvelle politique du matériel." L.A. 1347 (Nov. 24, 1951), 6.

253. ________. "L'encadrement de notre armée de l'air." L.A. 1352 (Dec. 29, 1951), 6.

254. ________. "La politique aérienne française: le choix des matériels." L.A. 1371 (May 10, 1952), 6.

255. ________. "La politique aérienne française: le bilan de cinq années." L.A. 1372 (May 17, 1952), 6-7.

256. ________. "Il faut que le ciel au-dessus de notre sol soit interdit à l'aviation adverse." L.A. 1322 (June 2, 1951), 6.

257. ________. "Sur le chemin de la raison." L.A. 1379 (July 5, 1952), 6.

258. ________. "Pourquoi n'avons-nous pas l'armée de l'air que nous voudrions et qu'il nous faut avoir." L.A. 1397 (Nov. 8, 1952), 6.

259. ________. "Pour un bilan véridique." L.A. 1434 (July 25, 1953), 6-7.

260. ________. "Vers un retour aux temps périmés de la 12e direction: sous la coupe de l'armée de terre?" L.A. 1435 (Aug. 22, 1953), 6.

261. ________. "L'aviation de transport." R.D.N. Oct., 1953, 296-313.

262. Germain, Commandant Louis. "L'armée de l'air en Algérie en 1959." F.A.F. 157 (Mar., 1960), 499-504.

263. Glavany, Colonel and Commandant Enault. "Le Jaguar." F.A.F. 237 (June, 1967), 823-828.

264. Glavany, Colonel and Commandant Brylinski. "Le Mirage F.1." F.A.F. 248 (June, 1968), 931-940.

265. Glavany, Général R. "Petite histoire des prototypes militaires français, 1945-1973." A.A. 286 (July, 1973), 64-69.

266. Gleizes, Commandant. "L'école de formation initiale du personnel navigant." F.A.F. 284 (Oct., 1971), 269-280.

267. "Le gouvernement doit, le plus tôt possible, adopter les mesures nécessaires pour refaire une grande armée de l'air." L.A. 1436 (Aug. 29, 1953), 6.

268. Grenet, Lieutenant-Colonel. "Le Mirage IIIE, avion des forces tactiques." F.A.F. 213 (Apr., 1965), 485-500.

269. Grigaut, Général C. "Les étapes d'une maturité, 1963-1985." A.A. 288 (Nov., 1973), 6-9.

270. Grimal, Lieutenant-Colonel. "Réflexions sur les écoles de l'armée de l'air." F.A.F. 4 (Jan., 1947), 472-489.

271. Groysillier, Colonel F. "Aperçus sur l'action des forces aériennes dans la guérilla." F.A.F. 104 (May, 1955), 801-820.

272. Guès, Lieutenant-Colonel. "De la rizière à la base aérienne." F.A.A. 87, supplement, 525-538.

273. Guhl, J.M. and A. Crosnier. L'ARMEE DE L'AIR EN INDOCHINE: TRANSPORT ET BOMBARDEMENT; 1945-1954. Paris: 1980.

274. Guyot, Général. "Réflexions sur l'aviation." F.A.A. 8 (May, 1947), 149-158.

275. Hartemann, Général. "Hier et demain dans le ciel." R.D.N. May, 1947, 613-627.

276. _________. "Les forces aériennes françaises dans une coalition." R.D.N. Aug., 1948, 184-200.

277. "L'hélicoptère lourd dans les opérations de maintien de l'ordre en Algérie." F.A.F. 150 (July, 1959), 33-44.

278. Hodeir, Marcellin. "Doctrine d'emploi et mission de l' armée de l'air, 1946-1948." R.H.A. 1982, Special number, 60-69.

279. Houard, Georges. "Que penser du chasseur VB-10?" L.A. 1140 (Dec. 6, 1947), 6.

280. ________. "Faut-il renoncer au Cormoran?" L.A. 1206 (Mar. 12, 1949), 12.

281. ________. "Ceux qui firent notre armée de l'air: le général Bouscat." L.A. 1311 (Mar. 17, 1951), 5-6.

282. ________. "Ceux qui firent notre armée de l'air: le général de Vitrolles." L.A. 1315 (Apr. 14, 1951), 5.

283. ________. "Une attaque, une contre-attaque." L.A. 1336 (Sept. 8, 1951), 6.

284. ________. "Formez en France des pilotes français." L.A. 1340 (Dec. 6, 1951), 6.

285. ________. "Des pistes mais peu d'avions." L.A. 1374 (May 31, 1952), 6.

286. ________. "Une conception française: le Potez 75." L.A. 1450 (Nov. 14, 1953), 6.

287. ________. "Une année d'aviation française." L.A. 1508 (Dec. 25, 1954), 7-10.

288. ________. "L'aviation française? On n'en veut plus." L.A. 1519 (Mar. 12, 1955), 2.

289. ________. "Un ministre de l'air? Nous devons l' avoir." L.A. 1525 (May 14, 1955), 7.

290. ________. "Des hélicoptères pour l'Afrique du Nord." L.A. 1571 (Mar. 10, 1956), 2.

291. ________. "Le Potez 75 en Afrique du Nord." L.A. 1572 (Mar. 17, 1956), 2.

292. "Le général Houdemon." L.A. 1798 (Nov. 5, 1960), 7.

293. "Il nous faut un avion d'appui." L.A. 1541 (Aug. 13, 1955), 7.

294. "En Indochine." F.A.F. 75 (Dec., 1952), 425-431.

295. "En Indochine: 450 avions en ligne." L.A. 1476 (May 15, 1954), 6.

296. "Instruction provisoire sur l'emploi des forces aériennes." F.A.F. 22 (July, 1948), 476-483.

297. Jacquard, Lieutenant-Colonel. "Réflexions sur la guerre aérienne en Extrême-Orient." F.A.F. 67 (Apr., 1952), 37-50.

298. Jacquot-Lacoussière, Lieutenant-Colonel. "Participation de l'armée de l'air à l' expérimentation de Reggan." F.A.F. 159 (May, 1960), 755-768.

299. "Le Jaguar: un avion différent." A.A. 286 (July, 1973), 48-57.

300. Jarry, Commandant. "Le renseignement aérien en Indochine." F.A.F. 125 (Apr., 1957), 689-708.

301. Jenlis, Ingénieur en chef de l'air. "Le Mirage III et sa fabrication." F.A.F. 147 (Apr., 1959), 605-620.

302. Kaeppelin, Thierry. "L'évacuation sanitaire en Afrique du Nord." F.A.F. 157 (Mar., 1960), 463-474.

303. Kerest, Frank. LES CANONS DU CIEL. Paris: 1960.

304. Kohl, Wilfred L. "The French Nuclear Force and Alliance Diplomacy, 1958-1967." Unpublished doctoral dissertation, political science, Columbia University, 1968.

305. Lacroix, Colonel R. "L'emploi des hélicoptères n' est-il justifié qu'en Algérie?" R.D.N. May, 1958, 834-845.

306. Lahitte, J.C. "Avec les combattants de l'air en Indochine: l'opération Artois." L.A. 1439 (Aug. 29, 1953), 7.

307. Lainey, Colonel. "De l'emploi des forces aériennes tactiques." F.A.F. 148 (May, 1959), 699-718.

308. Lauzin, Général Charles. "Opérations en Indochine." F.A.F. 102 (Mar., 1955), 449-487.

309. ________. "A propos de l'ALOA en Indochine." F.A.F. 108 (Oct., 1955), 577-586.

310. Lavoisier, Ingénieur en chef des travaux de l'air. "L' industrie aéronautique française face au marché commun." F.A.F. 149 (June, 1959), 1023-1054.

311. Lecerf, Lieutenant-Colonel. "Les bases d'une nouvelle doctrine aérienne." F.A.F. 85 (Oct., 1953), 183-200.

312. ________. "Réflexions sur la notion de supériorité aérienne." R.D.N. Dec., 1959, 618-626.

313. Léchères, Général C.F. "La France prépare son arme aérienne." INTERAVIA 5 (May, 1949), 260.

314. Lejay, Général. "L'aviation légère de l'armée de terre: le conflit terre-air en France." R.D.N. Aug.-Sept., 1956, 1070-1084.

315. Le Loher, Commandant. "Réflexions sur le corps des officiers mécaniciens." F.A.F. 249 (July, 1968), 67-96.

316. Lemaire, Capitaine A. "Un an d'opérations aériennes en Indochine." F.A.F. 67 (Apr., 1952), 111-132.

317. Le Meur, Lieutenant-Colonel. "La maintenance du Mirage IV." F.A.F. 282 (July, 1971), 17-38.

318. Lenthéric, Ch. "Est-il possible de donner une nouvelle orientation à l'industrie aéronautique française?" F.A.F. 89 (Feb., 1954), 191-198.

319. Léonard, Herbert. "Suez 1956: une aventure militaire, un échec politique." AVIATION MAGAZINE INTERNATIONAL 765 (Nov. 1, 1979) 82-87; 766 (Nov. 15, 1979), 78-83; 767 (Dec. 1, 1979), 79-85.

320. Leuba, André. "Ceux qui firent notre armée de l'air: le général Chassin." L.A. 1314 (Apr. 7, 1951), 5.

321. ________. "Ceux qui firent notre armée de l'air: le général Pierre Bodet." L.A. 1315 (Apr. 21, 1951), 6-7.

322. ________. "Ceux qui firent notre armée de l'air: le général Jacquin." L.A. 1320 (May 19, 1951), 6.

323. ________. "Ceux qui firent notre armée de l'air: le général Murtin." L.A. 1322 (June 5, 1951), 6-7.

324. Leyvastre, Pierre. "Visite dans l'est au ler CATAC." L.A. 1496 (Oct. 2, 1954), 6.

325. "La licence du Vampire." L.A. 1207 (Mar. 19, 1949), 12.

326. Lieber, R.J. "The French Nuclear Force." INTERNATIONAL AFFAIRS 42 (1966), 421.

327. Litré, Colonel B. and Lieutenant-Colonel G. Solans. "La défense aérienne: une nécessité nationale servie par un concept original." R.D.N. June, 1979, 93-108.

328. Luthereau, Lieutenant-Colonel. "ALOA en Indochine." F.A.F. 108 (Oct., 1955), 573-576.

329. Madre, Général. "Du rôle des forces aériennes en Afrique occidentale." F.A.F. 150 (July, 1959), 89-96.

330. "Le magnifique bilan d'activité de nos forces aériennes d'Indochine." L.A. 1124 (Aug. 9, 1947), 1, 5.

331. Mansuy, Colonel. "Pour ou contre le ministère de l' air." F.A.F. 20 (May, 1948), 165-183.

332. Marchand, Yves. "1952, résultats encourageants chez nos constructeurs d'avions." F.A.F. 75 (Dec., 1952), 371-397.

333. Marchand, Yves. "Tendances contemporaines dans l' industrie aéronautique française." F.A.F. 105 (June, 1955), 1137-1160.

334. Marie, Pierre. "Les idées du général Montrelay." L.A. 1127 (Aug. 31, 1947), 4, 6; 1128 (Sept. 6, 1947), 4-5; 1129 (Sept. 13, 1947), 5-6.

335. Martelly, Commandant. "Les enseignements aériens de la campagne d'Indochine." F.A.F. 35 (Aug., 1949), 584-612.

336. Martin, Commandant. "Le recrutement et la formation des pilotes de l'armée de l'air." F.A.F. 202 (Apr., 1964), 534-538.

337. Marin, Général. "L'armée de l'air dans le contexte nucléaire." R.D.N. Oct., 1964, 1499-1517.

338. Martin, Général. "L'armée de l'air dans le système militaire français." F.A.F. 207 (Oct., 1964), 411-438.

339. ________. "Armée de l'air d'aujourd'hui." R.D.N. Jan., 1967, 5-19.

340. Masenello, E.J. "Dien Bien Phu...tout le monde descend." L.A. 1460 (Jan. 23, 1954), 6-7.

341. ________. "Avec les combattants de l'air en Indochine: évacuations sanitaires." L.A. 1475 (May 8, 1954), 6.

342. Maurin, Général Philippe. "L'armée de l'air: réalités et perspectives." R.D.N. June, 1969, 962-972.

343. Mayer, Lieutenant-Colonel. "Considérations sur le recrutement des officiers navigants de l'armée de l'air." F.A.F. 107 (Aug.-Sept., 1957), 306-318.

344. Mellina, Bruno. "Les hélicoptères dans l'armée de l' air, 1950-1975." A.A. 297 (Sept., 1975), 58-66; 298 (Oct., 1975), 56-67.

345. Menanteau, Capitaine. "Le bombardement en Algérie." F.A.F. 150 (July, 1959), 57-64.

346. Mendel, Wolf. DETERRENCE AND PERSUASION: FRENCH NUCLEAR ARMAMENT IN THE CONTEXT OF NATIONAL POLICY; 1945-1969. New York: 1970.

347. "Le Mirage F.1." A.A. 290 (Mar., 1974), 32-41.

348. "Mission d'appui en Siebel dans la région d'An-Khé." L.A. 1427 (June 6, 1953), 6-7.

349. Mitaux-Maurouard, Commandant. "La défense des bases aériennes d'Indochine." F.A.A. 87, supplement, 596-602.

350. Monjauze, Médecin-Colonel. "Les évacuations sanitaires aériennes en Algérie." F.A.F. 122 (Jan., 1957), 93-102.

351. Montrelay, Général. "Caractères, organisation et évolution des forces aériennes françaises." F.A.F. 13 (Oct., 1947), 5-12; 14 (Nov., 1947), 165-174.

352. ________. "La défense aérienne du territoire." R.D.N. June, 1949, 725-739.

353. Moreau, Jean. "Le problème général de défense nationale." L.A. 1522 (Apr. 2, 1955), 2-3.

354. M.U. "L'aviation dans la guerre d'Indochine." L.A. 1446 (Oct. 17, 1953), 6-7.

355. Narbonne, Roland de. "Le SO-4050 Vautour et ses possibilités." L.A. 1458 (Jan. 9, 1954), 4-5.

356. ________. "A Cambrai avec la 12e escadre de chasse." L.A. 1533 (June 18, 1955), 11.

357. ________. "L'école de l'air." L.A. 1562 (Jan. 15, 1956), 6-7.

358. ________. "Avec l'armée de l'air en Algérie." L.A. 1583 (June 2, 1956), 6; 1584 (June 9, 1956), 7; 1585 (June 16, 1956), 6-7.

359. ________. "La 33e escadre de reconnaissance." L.A. 1611 (Dec. 19, 1956), 5-6; 1612 (Dec. 22, 1956), 6-7.

360. Nicolas, Commandant Bernard. "L'entraînement aérien dans les unités opérationnelles." F.A.F. 280 (May, 1971), 493-508.

361. Noetinger, Jacques. HISTOIRE DE L'AERONAUTIQUE FRANCAISE. 3 vols. Paris: 1978-1984.

362. Noroy, Commandant. "Le STRIDA." F.A.F. 279 (Apr., 1971), 365-372.

363. "Un nouveau quadrimoteur voit le jour: le Bréguet 482." L.A. 1136 (Nov. 1, 1947), 1.

364. Pagniez, Yvonne. LES AILES FRACAISES AU COMBAT. Paris, Geneva: 1957.

365. Palu, Thérèse. CONVOYEUSE DE L'AIR. Paris: 1957.

366. "Panorama de l'armée de l'air au seuil de 1960." L.A. 1774 (Mar. 23, 1960), 6-7.

367. Paquier, Colonel Pierre. "Des plans ou des actes." L.A. 1221 (June 25, 1949), 5-6.

368. _________. "Tonkin 1947. Le pont de l'aviation dans l' opération Léa." F.A.F. 50 (Dec., 1950), 141-156.

369. Pellenc, Marcel. SOMMES-NOUS DEFENDUS DANS LES AIRS? LE REARMEMENT AERIEN. Paris: 1951.

370. _________. RAPPORT SUR LES SOCIETES NATIONALES DE CONSTRUCTIONS AERONAUTIQUES. Paris: 1948.

371. Peguignot, Commandant. "GMMTA et union française." F.A.F. 117 (July, 1956), 181-188.

372. Personne, Pascal. "Les effectifs de l'armée de l'air de 1946 à 1949: une approche critique." R.H.A. 1982, special number, 16-27.

373. Pesquiès-Courbier, Simone. "Le renouveau de la medicine aéronautique de 1945 à 1959." R.H.A. 1982, special number, 16-27.

374. Pichoud, Ingénieur en chef de l'armement. "Les missiles nucléaires." A.A. 312 (June, 1978), 14-20.

375. "La politique de M. Louis Christiaens, secrétaire d'état à l'air." L.A. 1659 (Dec. 14, 1957), 7, 18.

376. Pollock, J. "France's six thorns." CONTEMPORARY REVIEW, May, 1952, 252-262.

377. Potez, Henri. "Défense de l'avion léger d'appui." F.A.F. 92 (May, 1954), 677-681.

378. "Pour les missions de reconnaissance et d'appui-feu en Algérie: le North American T-28, remplaçant éventuel du T-6." L.A. 1747 (Oct. 17, 1959), 8.

379. Puget, Général A. "L'industrie aérospatiale française." F.A.A. 226 (June, 1966), 757-777.

380. "Quand au parlement on parle d'aviation." L.A. 1222 (July 2, 1949), 1-2, 4.

381. "Quand M. Christiaens fait le point de notre aviation." L.A. 1469 (Mar. 27, 1954), 6-7.

382. "Quand le parlement se penche sur quelques problèmes." L.A. 1471 (Apr. 10, 1954), 6-7.

383. "Quelques aspects de la politique aéronautique française d'après les déclarations du secrétaire d'Etat à l'air." L.A. 1202 (Feb. 12, 1949), 1, 6.

384. "Quelques aspects du problème aérien." L.A. 1459 (Jan. 16, 1954), 6-7.

385. "Que serait la bataille?" R.D.N. Apr., 1950, 375-387.

386. "La question de la SNCAC et celle du Cormoran." L.A. 1223 (July 9, 1949), 1-2, 4.

387. Rafat, Commandant. "Contribution de l'armée de l'air à la mise en valeur de l'AOF." F.A.F. 106 (July, 1955), 28-40.

388. Ravaud, Jean. "L'industrie aérospatiale française." R.D.N. June, 1979, 25-40.

389. Raynaud-Karner. "Le pauvre budget des prototyes." L.A. 1266 (May 6, 1950), 6.

390. R.C. "L'évacuation de Na-San." L.A. 1441 (Sept. 12, 1953), 4-5.

391. "Le recrutement et la formation des personnels navigant et spécialisé." L.A. 1336 (Sept. 8, 1951), 6.

392. "Reims, 3e escadre de chasse." L.A. 1613 (Dec. 29, 1956), 8-9.

393. Riche, Jean-Marie. "La transformation de l'armée de l' air." L.A. 1906 (Dec. 28, 1962), 1.

394. Riffard, Marcel and Jean Romeyer. "Le développement des forces aériennes semble ralenti par un défaut de prévisions." L.A. 1561 (Dec. 31, 1955), 15, 17.

395. Riley, Colonel Daniel. "French Helicopter Operations in Algeria." MARINE CORPS GAZETTE, Feb., 1958, 15-18.

396. Robineau, Général Lucien. "L'adaptation de l'armée de l'air aux armes nouvelles." COLLOQUE: DE L' AERONAUTIQUE A L'ESPACE: 40 ANNEES DE DEVELOPPEMENT AEROSPATIAL FRANCAIS: 1945-1985. Paris: 1986, 57-70.

397. Rogé, Lieutenant-Colonel. "Armée de terre ou armée de l'air?" R.D.N. Feb., 1950, 217-219.

398. "Le rôle des hélicoptères dans la lutte contre la rébellion." L.A. 1683 (May 31, 1958), 13-14.

399. Romeyer, Jean. "Que valait le programme de 1945." L.A. 1303 (Jan. 20, 1951), 6.

400. ________. "La formule de Douhet reste valable." L.A. 1365 (Mar. 29, 1952), 6.

401. ________. "Peut-on concevoir un intercepteur moins coûteux et plus efficace?" L.A. 1375 (June 7, 1952), 6.

402. ________. "Quand les défaitistes du ciel accusent les circonstances." L.A. 1391 (Sept. 27, 1952), 6.

403. ________. "De la maîtrise de l'air à la guerre d' Indochine." L.A. 1403 (Dec. 20, 1952), 6.

404. ________. "Sur quelques aspects critiques du rapport du sénateur Pellenc." L.A. 1413 (Feb. 28, 1953), 6-7.

405. ________. "Après l'exercice Horace. Réalités et possibilités de notre DAT." L.A. 1428 (June 13, 1956), 6.

406. ________. "La place de l'air n'est pas la sienne. Pourquoi?" L.A. 1454 (Dec. 12, 1953), 6.

407. ________. "Les chefs de l'air sont trop souvent tenus à l'écart." L.A. 1461 (Jan. 30, 1954), 6.

408. ________. "Pourquoi il est nécessaire d'envoyer des avions en réaction en Indochine." L.A. 1479 (June 5, 1954), 6.

409. ________. "Le transport aérien à Dien Bien Phu." L.A. 1488 (Aug. 8, 1954), 6.

410. ________. "L'action de notre aéronavale à Dien Bien Phu fut admirable mais celle de notre armée de l' air aussi." L.A. 1490 (Aug. 21, 1954), 6-7.

411. ________. "En parlant avec le général Lauzin qui commanda l'air en Idochine: l'action du bombardement aérien." L.A. 1491 (Aug. 28, 1954), 6-7.

412. ________. "Le pont aérien de Dien Bien Phu: un fait unique dans l'histoire." L.A. 1502 (Nov. 13, 1954), 6-7; 1503 (Nov. 29, 1954), 6-7.

413. ________. "Comment réaliser notre transport aérien." L.A. 1515 (Feb. 12, 1955), 6-7.

414. ________. "Coup d'oeil sur le groupement des moyens militaires de transport aérien." L.A. 1517 (Feb. 26, 1955), 6; 1518 (Mar. 5, 1955), 6; 1519 (Mar. 12, 1955), 6; 1520 (Mar. 19, 1955), 6.

415. ________. "L'air dans une défense nationale efficace. Condition première: un chef responsable." L.A. 1521 (Mar. 26, 1955), 6.

416. ________. "L'armée de l'air et sa doctrine." L.A. 1525 (Apr. 23, 1955), 6-7.

417. ________. "Notre armée de l'air telle qu'elle est." L.A. 1542 (Aug. 20, 1955), 7.

418. ________. "L'éparpillement de l'air et ses conséquences: opinions parlementaires sur l' unité." L.A. 1544 (Sept. 3, 1955), 4.

419. ________. "Une DAT non pas inexistante mais insuffisante." L.A. 1545 (Sept. 10, 1955), 7.

420. ________. "Notre armée de l'air demain et après." L.A. 1548 (Aug. 27, 1955), 10.

421. ________. "Les forces aériennes dans le conflit algérien." L.A. 1570 (Apr. 28, 1956), 6-7; 1579 (May 5, 1976), 6-7.

422. ________. "En écoutant les aviateurs qui reviennent d' Algérie: l'avion type de l'union française." L.A. 1588 (July 7, 1956), 6.

423. ________. "Le GMMTA dans les opérations d'Algérie." L.A. 1591 (July 28, 1956), 7.

424. ________. "Le GMMTA dans l'union française." L.A. 1590 (July 21, 1956), 6.

425. ________. "Tâches et missions de l'armée de l'air." L.A. 1618 (Feb. 2, 1957), 7.

426. ________. "La querelle des aviations légères de terre et de l'air: ce n'est pas l'armée de l'air qui décide de ses missions mais la défense nationale." L.A. 1620 (Feb. 16, 1957), 6-7.

427. ________. "L'aviation légère de l'armée de l'air." L.A. 1622 (Mar. 2, 1957), 6-7.

428. ________. "L'armée de l'air en Afrique noire." L.A. 1624 (Mar. 16, 1957), 6; 1626 (Mar. 30, 1957), 6; 1627 (Apr. 6, 1957), 6-7; 1628 (Apr. 13, 1957), 6.

429. ________. "L'avion, élément capital de la sécurité du Sahara." L.A. 1667 (Feb. 8, 1958), 6.

430. ________. "La pensée militaire d'aujourd'hui: le général Bouvard et les hautes études." L.A. 1702 (Nov. 8, 1958), 4.

431. ________. "Ce que seront pour 1959 les crédits de l' aviation militaire: un budget, amorce de programme." L.A. 1715 (Feb. 7, 1959), 7.

432. ________. "L'engin de demain ne supprime pas l'avion: le besoin de reconnaissance reste entier." L.A. 1717 (Feb. 21, 1959), 7.

433. ________. "Une intéressante expérience française en Algérie: l'hélicoptère armé." L.A. 1721 (Mar. 21, 1959), 6.

434. ________. "L'armée de l'air dans une guerre subversive." L.A. 1729 (May 16, 1959), 9.

435. ________. "Un redoutable aspect de la future loi-programme: plus d'avions pilotés au-delà des Mirage." L.A. 1737 (July 11, 1959), 7.

436. ________. "Vers un déclin de notre puissance aérienne?" L.A. 1754 (Dec. 5, 1959), 6-7.

437. ________. "Composition de l'aviation de l'OTAN." L.A. 1961 (Jan. 23, 1960), 6-7.

438. ________. "La mission de présence." L.A. 1763 (Feb. 6, 1960), 6.

439. ________. "Justification et rôle de l'ALAT." L.A. 1766 (Feb. 27, 1960), 6.

440. ________. "Alors que notre industrie aéronautique ne cesse de s'affirmer: n'est-il pas possible de donner des avions français à nos combattants?" L.A. 1769 (Mar. 19, 1960), 6-7.

441. ________. "Quand l'armée de l'air s'équipe aux surplus américains: que penser de l'opération T-28?" L.A. 1771 (Apr. 2, 1960), 6.

442. ________. "L'ALAT en quête d'un programme: sommes-nous assez riches pour avoir deux aviations?" L.A. 1775 (Apr. 30, 1960), 6-7.

443. ________. "Faibles lueurs sur une politique." L.A. 1777 (May 14, 1960), 7.

444. ________. "Sur l'emploi de l'arme aérienne en opérations de maintien de l'ordre." L.A. 1786 (July 16, 1960), 6-7; 1787 (July 23, 1960), 6-7; 1788 (July 30, 1960), 6-7.

445. ________. "Loi-programme et logistique aérienne." L.A. 1789 (Aug. 6, 1960), 9.

446. ________. "La commission de défense nationale saisie de la loi-programme." L.A. 1793 (Oct. 1, 1960), 6.

447. ________. "Le ler CATAC et la détection radar intégrée." L.A. 1795 (Oct. 15, 1960), 6.

448. ________. "Les éléments essentiels de la loi-programme." L.A. 1798 (Nov. 5, 1960), 7.

449. ________. "L'aviation légère de l'armée de l'air." L.A. 1800 (Nov. 19, 1960), 6.

450. ________. "L'armée de l'air et ses crédits pour 1961." L.A. 1801 (Nov. 26, 1960), 6-7.

451. ________. "L'armée de l'air au Sahara." L.A. 1803 (Dec. 9, 1960), 7.

452. ________. "Ce que présente la loi de programme militaire." L.A. 1804 (Dec. 16, 1960), 7.

453. ________. "L'ALAT: l'armée de terre en quête de mobilité." L.A. 1808 (Jan. 13, 1961), 7.

454. ________. "L'armée de l'air dans la défense nouvelle." L.A. 1821 (Apr. 14, 1961), 7.

455. ________. "Ecoles de l'air: le personnel navigant." L.A. 1826 (May 19, 1961), 5; 1830 (June 16, 1961), 7.

456. ________. "L'armée de l'air et la défense aérienne." L.A. 1830 (June 13, 1961), 7.

457. ________. "Ecoles de l'air: personnels non navigants." L.A. 1831 (June 23, 1961), 7.

458. ________. "Ecoles de l'air: personnels officiers." L.A. 1832 (June 30, 1961), 7.

459. ________. "Le CIB: un creuset pour les équipages de la flotte stratégique." L.A. 1874 (Apr. 20, 1962), 7.

460. ________. "L'armée de l'air et la coopération." L.A. 1894 (Oct. 5, 1962), 4; 1895 (Oct. 12, 1962), 4.

461. ________. "Le COTAM et sa nouvelle physionomie." L.A. 1898 (Nov. 2, 1962), 7.

462. ________. "L'armée de l'air dans la défense opérationnelle du territoire." L.A. 1903 (Dec. 7, 1962), 4; 1904 (Dec. 14, 1962), 7.

463. Rosez, P. "Le personnel de l'Armée de l'air (novembre 1942-juillet 1950) Recherche d'une identité." Unpublished mémoire de maîtrise, Université de Paris X-Nanterre, 1982.

464. Rougeron, Camille. "Premières réflexions sur Dien Bien Phu." R.D.N. July, 1954, 20-31.

465. SABOTAGE DES AILES FRACAISES. Paris: 1960.

466. Sacco, Philippe. "L'armée de l'air en Indochine." A.A. 304 (Jan.-Feb., 1977), 48-59; 305 (Mar.-Apr., 1977), 52-63; 306 (May-June, 1977), 68-75; 307 (July-Aug., 1977), 74-79.

467. Saint-Cricq, Général M. "Les forces aériennes et les missions de défense." R.D.N. (June, 1979), 17-24.

468. Salvan, Lieutenant-Colonel. "Contrôle des opérations de défense aérienne et automatisation." F.A.F. 278 (Mar., 1971), 289-294.

469. Sauvaigo, Commandant. "Réflexions sur la défense aérienne." F.A.F. 131 (Nov., 1957), 579-594.

470. Savel, Lieutenant-Colonel. "La carrière d'officier de l'air." F.A.F. 32 (May, 1949), 211-227.

471. Schweisguth, Etienne, Mariette Sineau and Françoise Subileau. TECHNICIENS EN UNIFORME: LES SOUS-OFFICIERS DE L'ARMEE DE L'AIR ET DE LA MARINE. Paris: 1979.

472. "La sélection des pilotes de l'armée de l'air." L.A. 1692 (Feb. 2, 1958), 16-17.

473. "Le Sénat devant les crédits de l'armée de l'air." L.A. 1953 (Nov. 24, 1961), 7.

474. Servain, Jacques. "Une erreur capitale dans la défense du territoire." L.A. 1381 (July 19, 1952), 6.

475. _________. "Politique de prestige et politique nationale." L.A. 1411 (Feb. 14, 1953), 7.

476. _________. "Le drame de l'aviation de transport en Indochine: pas d'avions et pas d'équipements." L.A. 1424 (May 16, 1953), 6.

477. Siffre, Capitaine. "Le C-135F ravitailleur en vol des forces aériennes stratégiques." F.A.F. 270 (June, 1970), 701-702.

478. Simard, Général. "L'armée de l'air face aux problèmes de la réorganisation de la circulation aérienne en France." F.A.F. 228 (Aug.-Sept., 1966), 143-150.

479. Simard, Général. "L'armée de l'air et ses réserves." F.A.F. 278 (Mar., 1971), 237-254.

480. "Similitudes dramatiques dans l'histoire de l'armée de l'air." L.A. 1310 (Mar. 10, 1951), 6-7.

481. Simonpieri, Lieutenant-Colonel. "Le transall C-160." F.A.F. 280 (May, 1971), 519-530.

482. Stehlin, Général Paul. LA FORCE D'ILLUSION. Paris: 1972.

483. ________. LA FRANCE DESARMEE. Paris: 1974.

484. ________. "Les conséquences de la forme nucléaire de la guerre sur l'organisation de la défense nationale." F.A.F. 134 (Feb., 1958), 143-166; 135 (Mar., 1958), 301-328.

485. Teissier, Lieutenant-Colonel Yves. "Force de frappe et loi programme." F.A.F. 167 (Feb., 1961), 211-226.

486. Thorez, Maurice, André Marty and Charles Tillon. OU EN EST L'AVIATION FRANCAISE? Paris: 1948.

487. Traineau, Commandant. "Problèmes posés au COTAM pour l'exploitation des lignes régulières à longue distance." F.A.F. 242 (Dec., 1967), 843-859.

488. "Trident Blends Simplicity, Performance." AVIATION WEEKLY. Aug. 19, 1957, 51-53.

489. Tricaud, Capitaine. "L'aviation de liaison en Algérie." F.A.F. 150 (July, 1959), 65-72.

490. Turner, William H. "Another View of the Stalingrad and Dien Bien Phu Airlifts." A.H. (Sept., 1971), 135-137.

491. Valin, Martial. "Réflexions sur l'air intégral." R.D.N. May, 1949, 637-647.

492. "La véritable histoire du Cormoran." L.A. 1230 (Aug. 27, 1949), 5-6.

493. "Le vote du plan quinquennal." L.A. 1280 (Aug. 12, 1950), 1, 4-5.

494. X X X. "Rien ne justifie le doute." L.A. 1359 (Feb. 16, 1952), 7.

495. ________. "L'hélicoptère de combat dans la guérilla." L.A. 1570 (Mar. 3, 1956), 6-7.

496. ________. "Avec les hélicoptères du colonel Brunet dans le ciel d'Oranie." L.A. 1615 (Jan. 12, 1957), 5-6.

497. Z Z. "L'aviation légère dans la guerilla algérienne." L.A. 1576 (Apr. 14, 1956), 6.

498. Z Z Z. "Les engins spéciaux et l'armée de l'air." L.A. 1670 (Mar. 1, 1958), 14-15.